Partnering in the Construction Industry

Code of Practice for Strategic Collaborative Working

Partnering in the Construction Industry

Code of Practice for Strategic Collaborative Working

John Bennett and Sarah Peace
With the Chartered Institute of Building

Amsterdam • Boston • Heidelberg • London • New York • Oxford
Paris • San Diego • San Francisco • Singapore • Sydney • Tokyo
Butterworth-Heinemann is an imprint of Elsevier

Butterworth-Heinemann is an imprint of Elsevier
Linacre House, Jordan Hill, Oxford OX2 8DP
30 Corporate Drive, Burlington, MA 01803

First published 2006

British Library Cataloguing in Publication Data
A catalogue record for this book is available from the British Library

Library of Congress Cataloguing in Publication Data
A catalogue record for this book is available from the Library of Congress

ISBN 13: 978-0-7506-6498-1
ISBN 10: 0-7506-6498-3

For information on all Butterworth-Heinemann publications visit our website at www.books.elsevier.com

Printed and bound in Italy

Contents

Foreword

"Over the past decade, the construction industry has gained considerable experience of using partnering to complete a substantial number of construction projects successfully and to the satisfaction of their clients.

Following the recommendations in the Latham and Egan reports, it is encouraging to have seen many clients, consultants, contractors, subcontractors and specialists changing from the traditional adversarial relationships and discovering the benefits to be gained from a fully integrated way of working. In doing so they have been demonstrating the real scope for improvement in performance and laying the foundations for a genuinely world class and modern construction industry.

I recommend this new 'Code of Practice' as it brings together the best advice and the latest developments on the implementation of partnering. It is based on practical experience from individual projects and strategic alliances and uses real case studies to give a depth and breadth not found in any other publication.

The knowledge contained within this 'Code of Practice' is extensive and clearly set out. It is in a format that can be quickly and effectively used by all construction professionals in order to gain the benefits of past experience on all future construction projects.

It also forms a solid base on which further improvements in integrated working can be developed to ensure enhanced benefits and open the door to future generations of partnering."

Rt Hon Nick Raynsford MP
Deputy Chairman, Construction Industry Council

Preface

This code of practice explains what partnering is and why it delivers benefits. Designed for building owners, organizations responsible for infrastructure and construction industry firms, it provides detailed guidance about the actions that will help them begin to use project partnering effectively. It then describes actions that help clients, consultants, contractors and specialists steadily improve their joint performance further as they adopt strategic partnering and then develop it into strategic collaborative working. This journey enables them to match and eventually exceed today's best practice, providing massive benefits for everyone involved.

Partnering developed originally in North American manufacturing industries in the 1980s and 1990s as a response to Japan's strengths in key manufacturing industries. It was adopted by the construction industry first in the USA and then in the UK. These early uses of partnering in construction were directly influenced by research into Japanese construction practice. Now partnering is used extensively in western construction industries. It is supported by a substantial body of research that includes case studies of the successful use of partnering on individual projects and series of projects.

Partnering provides the basis for greater efficiency than older methods based directly on professional and craft practice and more recent methods based on project management. This is because partnering treats project teams as networks of work teams guided by well-developed communication links that include feedback systems. These characteristics define what science calls self-organizing networks. Modern science sees self-organizing networks as the most effective form of organization for all living things including human organizations.

The scientific explanation for the efficiency delivered by partnering was not recognized when partnering was first adopted by the UK construction industry in the mid-1990s. Nevertheless, early ideas, described in the authors' influential 1995 report *Trusting the Team*, are consistent with the science.

Trusting the Team essentially provides a theory of how partnering can be applied in UK construction based on three distinct research studies. These examined Japanese construction methods, partnering in the American construction industry, and the effects of long-term relationships in UK construction.

The theory represented by *Trusting the Team* was tested by the authors undertaking case studies of project teams using partnering guided by their report. The results show that project partnering can deliver benefits and these increase significantly when partnering is used over a series of projects. The research served to correct errors in the project partnering model described in *Trusting the Team* and identified a more developed model used by project teams experienced in partnering. This is described in the authors' 1998 report *The Seven Pillars of Partnering*.

Despite being out of date, *Trusting the Team* is still used to provide a basic guide to project partnering. This code of practice now brings that basic guidance up to date, taking account of the substantial body of subsequent research into partnering, key parts of which are listed in the Bibliography. Inevitably some of this research concentrates on weaknesses in partnering and the general sense of the criticism is taken into account to provide a robust basis for guidance on potential problems and weaknesses that project teams need to guard against.

In parallel to developments in theory and research, the UK construction industry has made substantial progress towards adopting partnering over the last ten years or so. It is used in 90% of the case studies currently published on the Constructing Excellence in the Built Environment's website. The Latham and Egan reports recommend the use of partnering although the Egan report, in concentrating on technological issues, calls it lean production. Both reports have been influential because, unlike earlier reports on the industry, the recommendations have been acted on by government, major clients and construction firms.

The UK construction industry now has the advantage of many guides to best practice. Those listed in the Bibliography are taken into account in writing this code of practice. The CIOB *Code of Practice for Project Management for Construction and Development* deserves special mention. It provides a robust description of the main stages in projects and best practice project management. It provides a sister publication to this code of practice and key points are summarized where they apply equally to partnering.

This code of practice begins with a strategic description of partnering for senior managers in client, consultant, contractor and specialist organizations. It explains that partnering involves initial costs and provides substantial time and cost savings that can be used to benefit their firms in whatever way they decide.

The main part then provides guidance for project teams about the actions they need to take in using partnering to improve their performance. It explains how partnering empowers designers, managers and specialists to do their best work. It does this by suggesting actions that encourage the development of technically competent work teams and by supporting them in using cooperative teamwork guided by feedback systems.

Partnering mirrors developments in other major industries where information technology and new forms of face-to-face meetings are revolutionizing work and business practices. We show that, in construction, project partnering can reduce costs by 30% and times by 40% compared to traditional approaches.

This code of practice explains that partnering is not a fixed way of working. It is shaped by project teams to fit the client's objectives and the kind of building or infrastructure they aim to produce. Also, partnering develops as project teams cooperate in achieving mutual objectives and performance improvements using agreed decision-making processes.

This code of practice describes the processes used in selecting the work teams that form project teams using partnering. It describes the organizational and commercial arrangements needed to give these work teams and their parent firms the confidence to aim for the maximum benefits from partnering.

It explains the central role of partnering workshops and provides detailed advice on their organization and running including the use of professional facilitators.

This code of practice explains how project teams using partnering lay the foundations for meeting agreed objectives by planning the design, construction and completion of each major stage of their work as an integrated system. It explains the care needed to ensure that work teams understand their work and its relationship to the rest of the project. It explains the need to reinforce cooperative teamwork and foster open and effective communication. It explains how partnering encourages project teams to search relentlessly for more efficient and effective ways of working. It explains how partnering encourages project teams to capture lessons for use on future projects.

This code of practice also provides detailed guidance for firms on internal partnering. This includes the actions needed to ensure their staff are skilled and experienced in using partnering. The actions often require firms to make fundamental changes to their structure and policies to fully support partnering. Managers need to be more involved in external than internal communications. Work teams have to be given the authority to make decisions and take actions as part of project teams. Management hierarchies become streamlined and less important. Policies become flexible as they are shaped to fit the needs of individual projects.

The final chapter of the main part of the code of practice, Chapter 6, begins by describing strategic partnering. This means a group of firms using partnering to improve their joint performance long term over a series of projects. In either case it can reduce project costs by 40% and times by 50% compared to traditional approaches.

Strategic partnering in most cases is based on the work of one major client but is also used by groups of consultants, contractors and specialists who have worked together successfully. Whatever the basis, the arrangement is led by a strategic team of senior managers from all the firms involved. They develop an explicit strategy that describes the type of buildings or infrastructure and services their joint organization will produce and market.

This code of practice provides detailed guidance on the way partner firms are chosen to provide the technological skills and business strengths to put the strategy into effect. It explains how financial arrangements can be designed to encourage partnering; and how the firms' organizations and processes are integrated. It provides guidance on making project processes ever more efficient by standardizing on best practice and actively searching for improvements outside of individual projects. It explains how the strategic team can ensure that key aspects of performance are measured to give senior managers in all the firms objective information about the benefits being delivered. It explains how the results are used to set targets for improvement and provide the basis for feedback systems.

Chapter 6 then describes the leading edge of current practice, which it calls strategic collaborative working. This takes two distinct forms. First, there are groups of construction firms that combine creative design with the technologies needed to produce highly individual buildings and infrastructure. They are skilled at satisfying the distinctive needs of many different clients ranging from rich individuals to global corporations. Second, there are groups of construction firms specializing in the total construction cycle of development, production and use for specific types of building or infrastructure. They provide sophisticated customer support services and market the total package under a brand name. They give potential customers a clear image of what to expect when they buy a new constructed facility. The code of practice explains how totally professional marketing is being used to create long-term sustainable businesses by building customer loyalty. In other words this part of construction is becoming a modern consumer product industry.

These two very significant developments allow the leading edge of construction to stand comparison with other leading twenty-first-century manufacturing industries. In their most highly developed forms both kinds of strategic collaborative working can reduce costs by 50% and times by 80% compared to traditional approaches.

The code of practice provides the most detailed and authoritative guidance on the actions that clients and construction firms take to achieve these levels of benefits. No one involved in any way with construction projects should risk being without it.

John Bennett
Sarah Peace

Acknowledgements

This *Code of Practice* has been written with help and advice from an editorial working group set up by the Chartered Institute of Building (CIOB). I am pleased to acknowledge the valuable contribution made by this group that comprised the following members:

Alan Crane	Chairman
Gavin Maxwell-Hart	Institution of Civil Engineers
John Campbell	Royal Institute of British Architects
Saleem Akram	Chartered Institute of Building
Ian Caldwell	Chartered Institute of Building
John Douglas	Englemere Limited
Rosemary Elder	Chartered Institute of Building
Barry Jones	Chartered Institute of Building
David Woolven	Chartered Institute of Building

Additional endorsements were provided by:

Rodger Evans	Constructing Excellence in the Built Environment
Judith Harrison	The Housing Forum
Adesh Jain	International Project Management Association
Jack Pringle	Royal Institute of British Architects
Steve Waite	Mansell
Martin Winn	Chartered Institute of Housing
Andrew Wolstenholme	BAA

Case studies are used throughout to provide practical examples. These are all taken from Constructing Excellence in the Built Environment's published case studies. I am pleased to acknowledge the help provided by Constructing Excellence in the Built Environment in identifying the case studies that illustrate best practice partnering. I particularly acknowledge the help provided in this way by Constructing Excellence in the Built Enviroment's Rodger Evans and Anna Yianoullou.

I would also like to acknowledge the help, encouragement and advice provided by Alex Hollingsworth who is the authors' main contact with the publishers, Elsevier.

Finally, I would like to thank on behalf of the CIOB both Professor John Bennett and Dr Sarah Peace for writing this book.

Professor Roger Flanagan FCIOB
Senior Vice-President
CIOB

How to use this code of practice

This code of practice is designed to be used flexibly by a variety of readers. It is arranged in sections that each provide specific guidance. Each section begins with a short executive summary printed in colour. The sections are arranged in chapters that each deal with a major aspect of partnering. The chapters are arranged in three main sections as follows.

Section 1 is in purple. It comprises Chapter 1, which is for senior managers in client and construction organizations who are thinking about using partnering for a construction project. It describes partnering including its costs and benefits and explains why projects using partnering are more efficient than those using other approaches.

Section 2, which provides the main body of this code of practice, is in brown. Chapters 2, 3 and 4 provide practical guidance for clients and construction firms using partnering on building and infrastructure projects. Chapter 5 describes internal partnering, which means the organizational arrangements that individual firms need to make to get the most from using partnering. Chapter 6 provides practical guidance for clients and construction firms using partnering strategically over a series of projects.

Section 3 is in blue. It provides descriptions of techniques and checklists used by leading practice. These can be read in isolation but Chapters 2 to 6 provide explicit cross references to the techniques and checklists. A PDF file of this section is available from the companion site to this book (http://books.elsevier.com/companions/0750664983).

The executive summaries at the start of every section allow busy readers to get an overview of the guidance provided in ten to fifteen minutes. The executive summaries also provide a guide for all readers to help them find the specific guidance they need.

Selected points from case studies of best practice published by Constructing Excellence in the Built Environment (between 1998 and 2004) are used to illustrate important points. This material is on separate coloured pages. It includes reference numbers that enable readers to identify the full published case studies on Constructing Excellence in the Built Environment's website (www.constructing excellence.org.uk).

Finally the Bibliography lists further reading about partnering, the reasons why it provides an efficient way of working, and techniques used in leading practice.

1

Project partnering explained

Project Partnering Defined

Project partnering is a set of actions taken by the work teams that form a project team to help them cooperate in improving their joint performance.

Specific actions are agreed by the project team taking account of the project's key characteristics, and their own experience and normal performance. The choice of actions is guided by a structured discussion of mutual objectives, decision-making processes, performance improvements and feedback.

1.1 Introduction

Project partnering is a set of actions that helps project teams improve their performance. It involves initial costs and provides substantial benefits. It is not a fixed way of working; it develops as project teams cooperate in finding the most effective ways of achieving agreed objectives.

Partnering is the most efficient way of undertaking all kinds of construction work including new buildings and infrastructure, alterations, refurbishment and maintenance. It provides more benefits than older, more established approaches.

This chapter provides an overview of partnering and its costs and benefits. It is intended for senior managers contemplating a new construction project who want to know how to get the best possible value for their investment. It is also aimed at senior managers in the construction industry, including consultants, contractors and specialists, as they develop strategies for improving their firm's performance and profitability.

Most published descriptions of partnering describe one version of current best practice. This can be misleading because partnering does not mean one single fixed way of working. It develops as people work together. The approach used on any given project is chosen by the client and project team, taking account of their experience of partnering, the nature of the project and the client's objectives. As a result some teams use partnering tentatively, others apply many of the features of published best practice, while a few have taken the ideas further to develop remarkable levels of efficiency.

This code of practice deals with the complexity of practice by first describing a straightforward approach to project partnering, acknowledging that it takes time to fully establish even that level of efficiency. This is discussed in the first four chapters of Section 2 of this code of practice. Then Chapter 6 describes how partnering is taken further by leading clients, consultants, contractors and specialists who use strategic partnering to work together long term. As the benefits grow some groups use strategic collaborative working to establish new and highly efficient businesses. Some specialize in original designs while others produce and market standardized constructed facilities backed up by sophisticated customer services. These exciting developments give construction the potential to become a genuinely modern industry comparable to any leading manufacturing industry.

1.2 The Challenge of Complexity

Teams undertaking construction projects face a task of remarkable complexity and difficulty.

Construction work should be exciting and rewarding for everyone involved. For many people it is, but there remain many cases where problems and disputes leave clients and people in the industry disappointed. This code of practice provides a set of practical actions that make it more likely that construction projects are successful and the people involved enjoy their work. Achieving these good outcomes is far from easy because modern buildings and infrastructure are complex products and construction is inherently difficult.

Modern buildings are complex, not in the sense of being highly sophisticated, but in bringing together many different technologies. Some building technologies are new and sophisticated but others are long established in trade practice. The design, manufacture and construction of even relatively simple buildings may involve close to a hundred different technologies and the most complex may need more than a thousand work teams with specialized skills and knowledge. No other human products give rise to these levels of complexity.

Modern infrastructure projects are very large and have to deal with a variety of environments and ground conditions that need different construction systems. New railways and roads are likely to involve cuttings, embankments, tunnels and bridges and include sophisticated information and communications systems. This variety of construction systems, often constructed in the midst of a very busy environment, makes large infrastructure projects complex.

The inherent complexity of buildings and infrastructure has caused the industry to fragment into thousands of small, specialist firms. As a result project teams comprise many individual work teams. They face a task made even more challenging by external complexity that arises because virtually every modern organization has an interest in buildings and infrastructure.

Most human activities take place in buildings and depend on the infrastructure that links them. Government at all levels regulates where, when and how new buildings and infrastructure can be produced. Many non-governmental agencies take an interest in the location and performance of buildings and infrastructure and the ways they are used. Special interest groups campaign for or against new buildings and infrastructure and take a deep interest in changes to the built environment. Private organizations make decisions that influence their own buildings and those contemplated or commissioned by neighbours. Everyone has an interest in proposals relating to buildings or infrastructure near their own home. Few other human products give rise to these levels of interest, opposition and support.

1.3 Demands for Efficiency

Despite the inherent difficulty of construction projects, clients rightly expect the industry to work efficiently.

Despite the inherent difficulty of the task, clients have every right to expect buildings and infrastructure to meet all their functional requirements, have low life-cycle costs and be produced efficiently. The challenges this provides for the construction industry are tough. Historically the UK construction industry has justifiably been criticized for failing to provide reliably good value for clients.

Things have changed over recent years and leading practice in the UK construction industry has made great strides in producing world-class buildings and infrastructure quickly and efficiently. This has been achieved by moving away from traditional practice, first by adopting project management techniques and more recently by using partnering. Understanding these changes will help clients make best use of the construction industry. It will also help the industry itself to improve yet further.

1.4 Construction Industry's Structure

Research identifies the construction industry's structure as a series of self-organizing networks. The basic elements are work teams, communication links and feedback that provide a robust basis for cooperative teamwork.

Research into partnering provides a distinctive picture of the construction industry. The basic unit in this new view of the industry is work teams. These are groups of people who specialize in specific design or construction activities together with the machines and systems needed to work effectively.

Work teams build links with other work teams. Some links arise because teams are employed by the same firm but the more significant links arise between teams working on the same project. These become very important when the same teams work together on a series of similar projects.

In these various ways the industry has become a network of work teams in which groups of teams establish links that enable them to work together effectively. Some of the links include feedback systems that guide the development of high levels of skill and competence. Repeated interactions between work teams guided by feedback give rise to specialized sectors of the construction industry.

Experienced clients work with the specialized sectors. Project teams are assembled by limited and carefully structured competition and negotiation from within the appropriate specialized sector. Experienced clients regularly employ a small number of consultants,

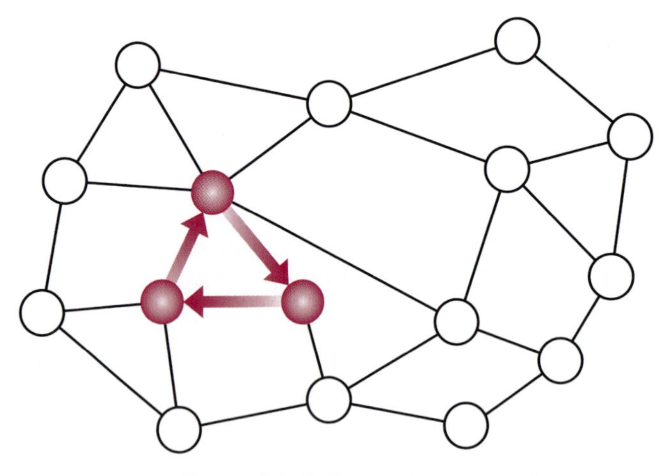

Figure 1.1 Self-organizing network

contractors and specialists who understand the technological and other challenges posed by the types of buildings and infrastructure they need. These sectors develop amazing levels of competence in building health centres, warehouses, superstores, service stations, offices, houses, sports stadia, roads, railways or bridges.

At the heart of these specialized sectors are feedback-driven clusters of work teams. It is entirely significant that they display all the characteristics that science generally has identified in controlled systems. Science now sees the whole planet as a richly interconnected network in which feedback gives parts the ability to survive and develop. This ability is called self-organization and science now identifies self-organizing networks as the most robust and effective form of organization for living things. Figure 1.1 shows a group of teams supported by feedback, which gives them the ability to work and develop within a wider and less controlled network.

Partnering is consistent with seeing the construction industry as a self-organizing network. It provides a set of actions that reinforce the natural grain of all effective living organizations and turns construction projects into efficient controlled systems. These characteristics explain why partnering is fundamentally more efficient than other ways of working. Figure 1.2 shows the key elements of controlled systems, all of which need attention from project teams.

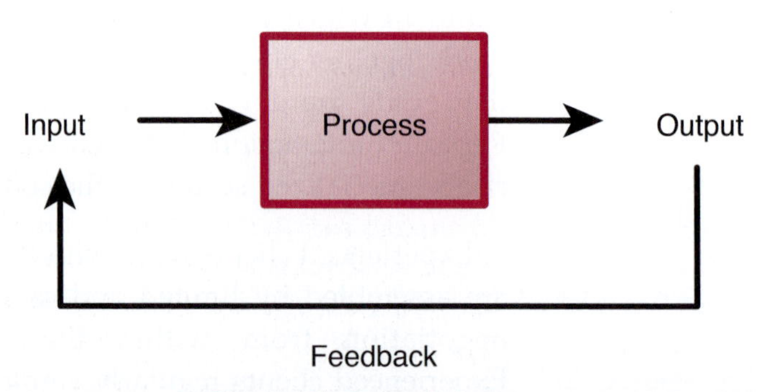

Figure 1.2 Controlled system

1.5 Traditional Practice

Traditional practice relied on independent firms brought together by competitive bids and tough contracts. Identifying the weaknesses of this approach helps explain how partnering provides benefits.

Traditionally the construction industry had a structure based on the perceived status of the various professions and trades. But it provided no explicit coordination or control. Consultants fiercely maintained their independence, contractors competed for work and specialists struggled to maintain the integrity of their skills and knowledge against market-driven demands for lower costs and faster delivery. Clients dealt with an industry that appeared chaotic by using competitive tenders and tough contracts to protect their own interests.

Project teams, assembled from work teams brought together often for the first time, relied on professional and trade practice to coordinate their work. The approach failed because it provided no overall direction, reducing everyone involved to defending their own interests. It ignored the need for well-developed links between workers that are the hallmark of effective teams. Despite these weaknesses, some clients are attracted by the simplicity of inviting competitive bids and, encouraged by advice from professionals with a vested interest in old ways of working, continue to use traditional methods. All too often they are sadly disappointed as they discover that claims, delays, defects and disputes make this an expensive and ineffective approach.

1.6 Project Management

Project management provides a better approach for construction. Cost, time and quality are controlled to achieve the client's objectives. Some designers claim to be inhibited by these management controls but in practice design outcomes tend to be better than on traditional projects where designers take the lead. Indeed many leading designers welcome having management issues dealt with by specialists so they can concentrate on design.

The construction industry has long recognized the failures of the traditional approach and in recent years has gone a long way towards resolving them by using project management. Best practice is well described in the CIOB's publication *Code of Practice for Project Management for Construction and Development*. This well-established code of practice, now in its third edition, describes the overall structure and processes created by project management techniques.

Project management improves the performance of project teams by creating a management role with strong links to the client and all the work teams. The additional costs of project managers and strong links are more than offset by greater efficiency as work teams benefit from being told how, where and when to work.

Specific client objectives sometimes force project managers to regard creativity and innovation as risky and this can cause a concentration on cost and time at the expense of quality. Designers and specialists often complain about these controls and claim they result in dull designs. Equally many leading designers welcome having a project manager deal with management issues, leaving them free to concentrate on design.

In practice management disciplines result in designs that on average are better than on traditional projects where designers take the lead. Design-led projects can produce outstanding designs but are equally likely to result in mediocre outcomes. Project management encourages consistently good design, and in doing so it may occasionally miss a masterpiece but it reliably avoids disastrously bad design.

1.7 Emergence of Partnering

Partnering empowers designers, managers and specialists to do their best work by establishing communication links and feedback systems. This mirrors developments in other major industries where information technology and highly developed forms of face-to-face meetings are revolutionizing work and business practices.

Partnering delivers significant improvements in performance by empowering designers and specialists to do their best work. Partnering grew out of revolutionary changes in other industries, notably the car industry. These other industries found ways of retaining the greater efficiency of work teams resulting from the use of management techniques whilst ruthlessly cutting management costs. They have been considerably helped in this so-called downsizing by rapid developments in information and communication technologies. The changes have radically altered the work of senior managers. They now concentrate on providing leadership, communicating internally and externally, and acting as coaches and mentors to their subordinates. UK construction began adopting these ideas in the 1990s largely in response to demands from major clients who could see the benefits in their own organizations.

Partnering in construction begins with very careful selection procedures. These rely on questionnaires, interviews and negotiations designed to ensure that the work teams forming a project team are competent and will work cooperatively. Price and cost play a minor role. The aim is to select an effective project team able to concentrate on doing its best work. It is possible to use competitive bids, if the client insists on this, without undermining the basis of partnering.

Partnering empowers designers and specialists to use cooperative teamwork in making their own decisions through networks of well-developed communication links. A project manager may be included in the team to ensure that quality, time and cost control systems are used effectively. As in other industries, information and communication technologies provide essential support. These modern developments often require firms to reorganize themselves internally to actively support work teams using partnering.

The changes deliver benefits relatively quickly in construction because they build on natural ways of working used throughout the industry. People generally choose to cooperate with others who make their work easier and more successful. Small builders use the same tradesmen, architects use the same consultants, site managers use the same specialist contractors because they work reliably and when there are problems, they help solve them. These natural ways of working are efficient and always have been. The industry's poor performance is caused by work teams being forced into an adversarial defence of their own interests by competitive tendering, tough contracts or outdated management ideas.

Partnering builds on these natural and efficient ways of working. Clients discuss their projects with consultants, contractors and specialists to agree the best ways of achieving agreed objectives. Designers look to specialist contractors and manufacturers as a wonderful source of new ideas and solutions to problems. Contractors integrate their supply chains. They are all helped in this by forms of contract that deal explicitly with cooperative teamwork.

The benefits produced by these developments have encouraged leading clients, consultants, contractors and specialists to work together on a long-term basis. This gives rise to strategic partnering, which is usually based on the work of one major client. It develops further into strategic collaborative working which is providing the basis for a genuinely modern construction industry that reliably delivers exceptional value for clients and robust profits for the construction firms involved.

Figure 1.3 illustrates the various approaches to construction projects in theoretical terms. The theoretical measures of the benefits delivered by partnering are based on research into the actions taken by experienced construction professionals on a range of projects (Bennett, 1997). It is entirely significant that this theoretical analysis is supported by data from construction projects given in Section 1.8.

1.8 Benefits of Partnering

The initial costs of establishing partnering are rapidly outweighed by the benefits, which include lower prices for clients, higher profits for consultants, contractors and specialists, faster completions, greater certainty and zero defects. Project partnering can reduce costs by 30% and time by 40%, while strategic collaborative working over a series of projects can reduce costs by 50% and time by 80%.

Research shows beyond reasonable doubt that, properly applied, partnering reduces the price clients pay for a given building. At the same time consultants, contractors and specialists earn better than normal profits and the industry's workforce find their work more rewarding in every sense.

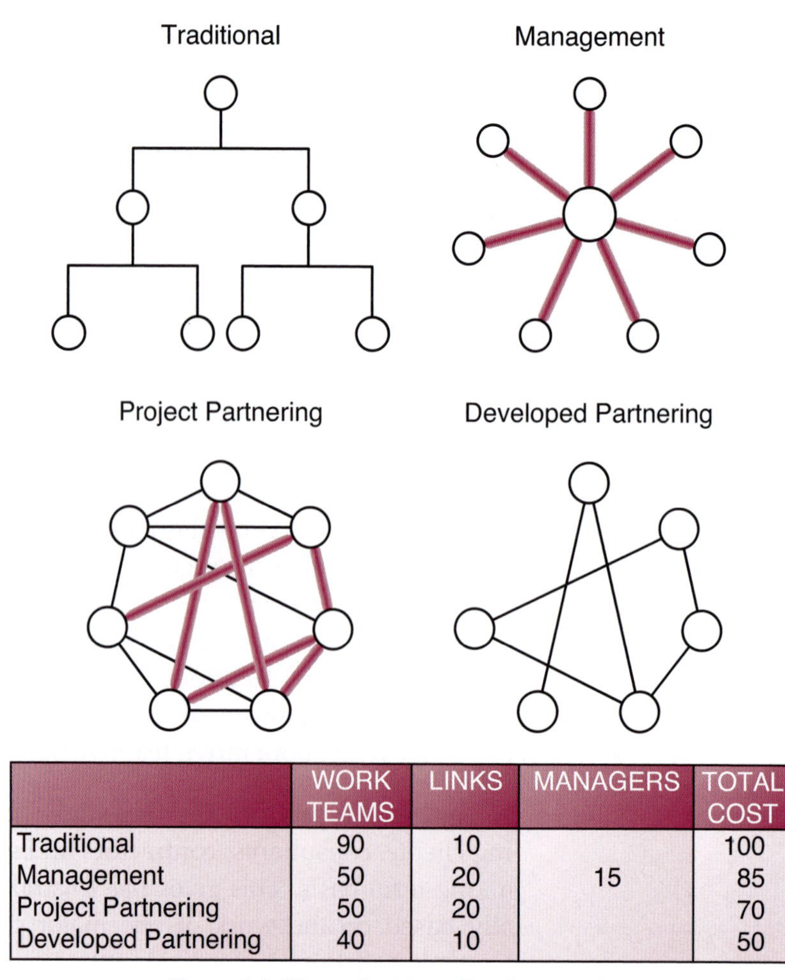

	WORK TEAMS	LINKS	MANAGERS	TOTAL COST
Traditional	90	10		100
Management	50	20	15	85
Project Partnering	50	20		70
Developed Partnering	40	10		50

Figure 1.3 Theoretical benefits of partnering

Case studies show that project teams partnering for the first time can deliver substantial benefits but it takes time and the experience of several projects for the full benefits to be realized. The benefits have been measured by comparing the performance of project teams using traditional and project management methods with those using partnering at various stages of development. The results in the box opposite come from the author's *The Seven Pillars of Partnering*. The data is consistent with many research studies into international construction performance including those listed in the Bibliography. The data is also consistent with performance improvements described in Constructing Excellence in the Built Environment's case studies.

Project partnering means a project team partnering on an individual project. Strategic partnering means firms supporting project teams in partnering over a series of projects. Given time and experience it develops into the most efficient forms of partnering, which we call strategic collaborative working.

The greater efficiency delivered by partnering is used by experienced clients to improve the performance of the end product, provide higher quality, more sophisticated controls, lower life-cycle costs, greater sustainability or other improvements of value to the client.

Performance Improvements over Traditional and Management Approaches by Project Teams Using Partnering Successfully

	Construction Costs	Construction Times
Traditional approaches	100	100
Management approaches	85	70
Project partnering	70	60
Strategic partnering	60	50
Strategic collaborative working	50	20

Note: a representative sample of projects using traditional approaches was used to establish the datum of 100 for construction costs and times. Samples of projects using management approaches and partnering at three distinct stages of development were used to establish the reductions in construction costs and times shown.

1.9 Criticisms of Partnering

Like all major changes, partnering provokes criticism from practitioners and academics. This provides a basis for a partnering checklist for senior managers.

Given the nature of the changes required to put partnering into effect, it is inevitable that some practitioners voice criticisms of the approach. It is equally inevitable that some academics respond to these criticisms by searching for problems and weaknesses. The following criticisms were identified by a review of the partnering literature.

- Organizations trying to establish a partnering culture for specific projects face severe problems when they have to use cut-throat competition to win other projects.
- Modern forms of decentralized decision-making undermine partnering as decisions by one department are contradicted elsewhere.
- Commercial realities that require firms to have alternative suppliers and many customers inhibit the development of deep partnering relationships.
- The open communication required by partnering is inhibited when one partner also works with another partner's competitors.
- Partnering relationships inhibit firms from developing more profitable new businesses.
- Teams responsible for individual projects achieve shallow forms of partnering because the approach takes time to develop.
- Forming teams from people who fit the partnering ideal excludes creative individuals, new ideas and distinctive skills.

- Powerful partners dictate terms and conditions to weaker partners who depend on them for future work and so cooperative teamwork is impossible.
- Senior managers retain detailed control so that work teams lack the freedom to become cooperative team players.
- Partnering is undermined by targets that focus on aspects of performance that are easy to measure.
- Partnering is undermined by targets that expect too much too soon.
- Partnering is undermined by targets that can be achieved only at the expense of those further down the supply chain.
- Attempts to standardize on the most efficient processes and designs undermine quality and value.
- Construction professionals only provide feedback that is directly relevant to their own firms' projects.
- Feedback is used in different ways at different levels in organizations and gets distorted, and important background information is lost.
- Partnering is undermined when commercial and organizational conditions change.
- Strategic collaborative working relationships too often mean that individual projects are sacrificed in the interests of long-term development.
- Some benefits attributed to partnering are equally well provided by different arrangements.

These various criticisms of partnering serve as a reminder that partnering is not an easy option. It is tough. It has to be worked at by everyone involved to achieve the full benefits. The lessons identified by this review of common criticisms of partnering provide a checklist for senior managers (see Chapter 7, checklist 5).

1.10 Costs of Partnering

The changes required to put project partnering into effect give rise to some initial costs for all the firms involved.

Partnering involves costs, which represent an initial investment that has to be met before the benefits emerge. The costs include time spent by senior managers in establishing the approach, careful team selection procedures, and training and partnering workshops.

These investments can be made gradually as the benefits emerge. It takes time for project teams to develop the abilities needed to use partnering effectively, so it makes sense to begin with small steps. Partnering can be developed by giving the same team a series of small projects. Some clients wanting to use partnering on a large project give the project team a small project first so they learn how to work together. If these arrangements are not possible, partnering can still be used by allowing time for the project team to discuss and agree how they will work.

1.11 The Client's Decision to Use Partnering

Investing in construction begins with key decisions about the building or infrastructure and the project team. These include balancing the initial time and resources required by partnering with the potential benefits. A decision to use partnering means selecting a project team in which all the members are willing to use cooperative teamwork.

The remainder of this chapter describes the situation facing clients new to partnering. However, clients experienced in using partnering successfully and wanting to take their approach are well advised to work through the basic steps to ensure that their decisions are well founded.

Construction projects normally arise from developments in clients' businesses. When it appears that a construction project may be needed, the project sponsor, who formally represents the client, should be appointed and given the support of an internal team. The team members should be selected carefully so they work together on the basis of cooperative teamwork. Between them they should provide a practical understanding of the relevant business issues, current trends and influences, industry norms and financial constraints. They need to understand how the business opportunity translates into physical requirements. Experienced clients may have this knowledge and experience in-house. Occasional clients may need to appoint consultants to provide advice on construction issues.

The internal team's first task is to define the need. In doing this, they should consider various ways of dealing with the new business situation. These include outsourcing work, reorganizing the use of existing facilities, extending existing spaces, leasing or buying an existing facility and commissioning a new building or infrastructure. If the best answer requires construction work, the internal team should produce a formal statement of the client's objectives.

The client's objectives should take account of all the main interests, including finance managers, facilities managers and users. Some organizations identify representative groups of customers to consult about changes to buildings or infrastructure. There are often other key interests including shareholders, neighbours, local authorities and trade unions. Wide consultation usually provides many ideas about how the building or infrastructure could help the organization be more successful. In all the discussions the internal team should take a wide view by considering the total life-cycle costs and environmental impacts.

Once the formal statement of the client's objectives is agreed, the client needs to make some key decisions about the project and how it should be run. These determine the nature of the building or infrastructure, the form of project organization to be used and whether to use partnering.

If they decide that partnering may be the best approach, the client needs to select consultants, contractors and specialists willing to work on the basis of cooperation. It is an advantage for the firms to be experienced in partnering but this is not always possible. It is vital that firms are technically competent but beyond that they should be clearly willing to take account of others' interests in agreeing decisions. This is particularly important for the firms that provide members of the project's core team. This is the small group of key individuals who with the project sponsor provide the project's overall direction.

Having selected the key members of the project team, the next action depends on the client's decision about partnering. Clients who have decided to use partnering can make arrangements for the first partnering workshop. Clients who want to give further consideration to partnering should hold a meeting with the project team to discuss how the project should be run. The meeting should take about half a day and aim at reaching a consensus about how they all want to work together. If the client with the project team agrees to use partnering, they should use the guidance given in the rest of this code of practice. Clients who decide not to use partnering will take other actions guided by the *Code of Practice for Project Management for Construction and Development*.

1.12 Actions by Construction Industry Firms

Consultants, contractors and specialists should take a positive and realistic attitude towards the use of partnering in discussing new projects with clients.

When consultants, contractors and specialists have an opportunity to discuss a new construction project with a client, they should provide information that helps the client make the initial key decisions. In doing this they should discuss partnering and their experience of working cooperatively. In any discussions about how the project will be run, they should take a positive attitude towards partnering but check that other people involved are genuinely prepared to work cooperatively. Any concerns about other parties' abilities or attitudes should be discussed before agreeing to use partnering. Everyone needs to be tough about their own interests and concerns. This may occasionally mean not being appointed and in most cases this means missing out on a bad project. Successful projects result from people being open and clear about their own interests and discussing them in a cooperative search for answers that give everyone what they need. This is how partnering works.

1.13 First Partnering Workshop

The client with the project team should prepare for and hold the first partnering workshop. It takes two days and uses professional facilitators to ensure the project's objectives and ways of working are considered fully and openly in a spirit of cooperation.

As soon as a client has decided to use partnering and appointed the key members of their project team, the first partnering workshop

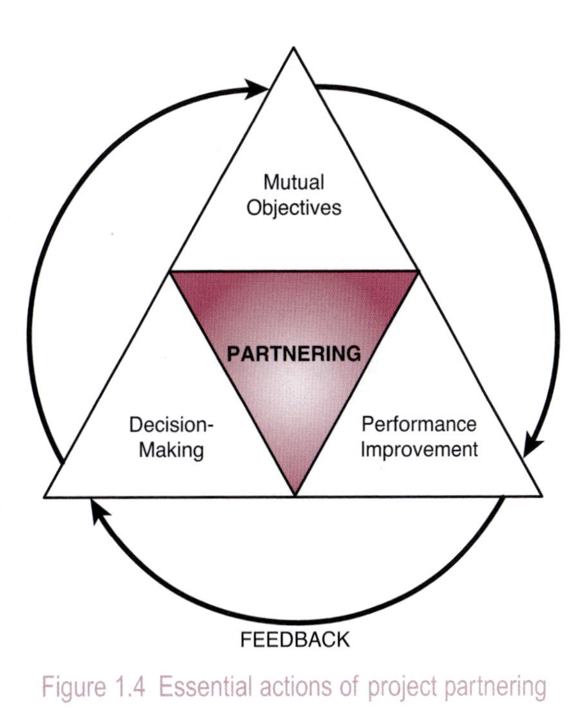

Figure 1.4 Essential actions of project partnering

should be held. This is where the essential features of partnering are tailored to the client's objectives and the project team's needs and concerns. Partnering workshops go well beyond the workshops used in value and risk management in shaping all the project inputs, processes and outputs and establishing the basis for cooperative teamwork.

All parties able to influence the outcome of the project should be present at the workshop. It normally takes two days and is held at a venue that takes everyone away from their normal working environment so they can concentrate on the project. It helps if it is run by a specialist partnering facilitator who will help people look beyond their own narrow advantage and realize they gain most by concentrating on the project's overall success.

The specific approach to partnering adopted by a project team is established at the first partnering workshop by considering the actions in Figure 1.4. The main features of these actions are described in the rest of this chapter to help senior managers understand why a first partnering workshop is being held and what it should achieve.

1.14 Mutual Objectives

When people cooperate in adopting a 'win–win' attitude they produce enough for everyone to have everything they reasonably want. In this spirit the first partnering workshop can agree mutual objectives that give the client a lower price and consultants, contractors and specialists higher profits.

Partnering accepts that firms look after their own interests. It requires a tough-minded recognition by clients that they will get what they

Mutual objectives may deal with many process- and product-related issues but the following should be considered.

- Value for money
- Guaranteed profits
- Reliable quality
- Fast construction
- Handover to owner on time
- Cost reduction
- Costs within agreed budget
- Operating and maintenance efficiency
- Improved efficiency for users
- Architectural quality
- A specific technical innovation
- Excellent site facilities
- Safe construction
- Shared risks
- Timely design information
- Shared use of computer systems
- Effective meetings
- Training in decision-making skills
- Training in management control systems
- No claims.

need only if consultants, contractors and specialists have a realistic opportunity to do good work and make reasonable profits. It requires an equally tough-minded recognition by consultants, contractors and specialists that they prosper best when clients get excellent value, good buildings or infrastructure and no hassle. This focus on mutual objectives gives expression to the idea that when people cooperate, they can produce more than enough to give everyone what they reasonably want. This is often described as a 'win–win' attitude in contrast to the traditional zero sum assumption that if one person gains, someone else must lose.

Clients should ensure that agreed mutual objectives take account of the interests of everyone affected by the project, so time and resources are not wasted on designs that will create problems. It may take time to deal with everyone's concerns. Inevitably the client, designers, managers, specialist contractors and manufacturers have different views about what constitutes success. And people tend to worry that in some way they will lose out if they cooperate in meeting other people's needs. Despite these perhaps natural reservations, experience shows that when project teams are brought together to discuss their individual interests they can find mutual objectives.

In agreeing mutual objectives it is essential to sort out the financial arrangements so that everyone gets a fair return in business terms. The worst situations in construction projects arise when any of the consultants or contractors are losing money. Contractual arrangements should ensure that none of the firms, if they contribute their best efforts, will lose out relative to the others. The essential equity of good value for clients and fair profits for consultants, contractors and specialists provides the platform on which partnering flourishes.

1.15 Decision-Making

The first partnering workshop agrees the decision-making processes needed to achieve the mutual objectives. They include procedures to ensure that problems are resolved quickly, in most cases by the people directly involved.

Construction projects bring together many work teams drawn from many different firms. They need to agree how decisions will be made. The aim should be to capture good ideas even when they come from unexpected sources. The air conditioning specialist may see the solution to a problem that is baffling the architect and engineers. A quantity surveyor may see the answer to a construction problem that is holding up work on site. Wide discussion may capture most of these inspired contributions but it takes time. Since most projects have to be completed quickly, a balance has to be found between accepting satisfactory answers that are readily available and continuing to search for something better.

The nature of the decision-making systems is directly influenced by whether the client needs the project to produce a standard answer or an original design. An important consequence of this choice is the amount of time the client and his staff will need to spend in making decisions. Original designs take more time but should result in buildings and infrastructure that support the users' activities and delight everyone who sees them. Standard answers make fewer calls on the client's time, are quicker and cheaper but impose more compromises on users and may look dull.

These issues are considered at the first workshop as the project team agree the information and communication systems they will use. They decide the quality, time and cost control systems they will use. They agree who will operate them and who will get the various outputs. They decide on the form and frequency of face-to-face meetings.

They consider the use of task forces, workshops, common project offices, social events and other ways of bringing teams closer together.

Whatever decision-making systems are agreed, they should include robust procedures to ensure that problems are resolved quickly in ways that encourage cooperative teamwork. This means most problems are resolved by the work teams directly involved. When a problem cannot be resolved in this way, it should be referred immediately to the project's core team and in exceptional cases to senior managers.

1.16 Performance Improvement

The first partnering workshop should agree a specific, measurable improvement and decide how it will be achieved. This may require a task force to develop the workshop's ideas further.

The whole point of partnering is to improve project teams' performance. Partnering that merely provides mutual objectives and agreed ways of making decisions will drift into inefficient ways of working. Partnering requires project teams to search for better answers.

Project teams new to partnering should aim at one modest improvement that all members of the team regard as important. A scattergun approach of aiming at several improvements at the same time usually leads to them all being missed. As experience of partnering grows, the scale and range of improvements will increase.

It is important that performance improvements in one area do not distract work teams from continuing to deliver their established normal performance in all other areas. This is an easy trap to fall into as attention is focused on the improvements, and quality elsewhere slips without anyone noticing. This is why partnering procedures give explicit attention to the constraints of achieving normal performance as well as delivering performance improvements.

There is controversy about the best ways of encouraging work teams to improve their performance. Many economists and traditional mangers argue that competition is the only reliable spur to improve performance. Unfortunately experience shows that competition in the construction industry can easily become cut-throat so bid prices, quality and safety are driven down to levels that are hopelessly inefficient. The outcomes include claims, disputes, defects, late completions and good firms being driven out of business.

Competition has a place in partnering when it encourages consultants, contractors and specialists to invest in training and innovation to improve their own performance. This can be achieved even when there are long-term relationships between firms. By having two, three or four options available for key relationships, all the partners are motivated to continuously improve their performance.

Benchmarking provides another weapon in the search for improved performance. Carefully researched information about best international practice is often used by experienced clients to guide the choice of targets. A good approach is to concentrate on whatever the client, consultants or contractors regard as their biggest problem.

There are advantages in project teams setting their own targets. When teams are given good information about the performance achieved by leading practice, they often set tougher targets than any they would accept from their managers.

Having agreed the performance improvements they will aim for, the best partnering teams try various ideas, continue with actions that work and change those that deliver no improvements. They often set up a task force to help find ways of meeting targets. This is a small group of people with relevant knowledge selected from within the project team and it may include external experts. Task forces should be given a short time to find an innovative answer that will deliver significant performance improvements.

The first partnering workshop should ensure that actions found to deliver improvements will be built into standards and procedures for the benefit of the current and future projects.

1.17 Feedback

Teams need to be guided by feedback about their own performance if they are to deliver the substantial benefits that partnering can provide.

Achieving performance improvements depends on project teams being provided with up-to-date and objectively measured feedback. Teams should measure their own performance and plot the results on control charts that show graphically how they are doing against their targets. Teams believe feedback they have produced themselves and use it to search for better ways of working. Feedback is most effective when it is expressed in positive terms. For example, quality should be measured by recording how often quality standards are achieved, not the number of failures.

Performance improves faster when successes are publicized and celebrated. It is vital that senior managers know when targets are being achieved and make a point of congratulating and rewarding the people involved. The rewards can be token but a dozen cans of lager presented at a light-hearted ceremony to the week's best work team can ensure that all the teams strive to be winners next week.

Failures must not be ignored. This is not to allocate blame, which is counter-productive. Failures should be used to guide teams in looking for robust answers to problems so that performance is back on target quickly. Some effective teams make a point of celebrating failures because they provide opportunities to find more effective ways of working.

When a failure arises, they have a party and then, with renewed enthusiasm, concentrate on finding a robust answer.

It is important that senior managers are kept up to date about improvements in performance. This is essential if they are to remain committed to partnering. At least some managers in most organizations take a pride in being highly competitive and are sceptical of the idea that the cooperative methods used in partnering can possibly be effective. Without regular, well-founded feedback on the performance improvements delivered by partnering, there is always a risk that adversarial methods will be reintroduced.

Feedback should flow from project to project. Too many innovative ideas are lost because construction generally has weak feedback systems. Lessons need to be captured so that good ideas are applied on future projects and problems and defects do not recur. Leading firms involved in partnering have developed standards and procedures that systematically capture best practice as it emerges from their projects. The feedback-based standards and procedures help all their project teams concentrate on efficient work. As Chapter 6 explains this is an essential element in using strategic partnering and strategic collaborative working successfully.

1.18 Maintaining Partnering throughout Projects

Progress towards mutual objectives and performance improvements using agreed decision-making processes is reinforced by partnering workshops throughout projects. A final partnering workshop captures lessons for use on future projects.

Best practice includes workshops throughout projects to review progress and if necessary change things agreed at the first partnering workshop. Change may be in response to the project going better than expected and the team realizing they can aim for bigger performance improvements. It is perhaps more common for projects to face problems. These should be discussed at a workshop, which if the problem is sufficiently serious, should be specially. The workshop should look for and agree actions that deal with persistent problems once and for all. Partnering is action-oriented and dealing with problems quickly is central to its success.

A final workshop is used to identify good ideas and lessons identified during the project so they can be recorded and made available for use on future projects.

All partnering workshops are organized in a similar way to the first partnering workshop and should be taken just as seriously. Partnering is an ongoing activity guided by workshops. The potential benefits are large and they are earned by concentrating on and continually reinforcing cooperative teamwork. This code of practice provides detailed guidance on the actions needed to adopt these most efficient ways of working.

Trust more effective than contract

Case Study Reference: M4i 121

A training facility in Merseyside shows that partnering is far more effective than tough contracts at pushing quality up and driving cost down.

The building resulted from an innovative public/private joint venture, Partnership for Learning, set up to provide hard and soft skills training for local industries. Supporters include many SMEs and big companies, including GlaxoSmithKline and Ford Jaguar.

The brief for the building set tough standards including requiring it to have low energy use and a minimal impact on the environment. This was achieved thanks to innovative ideas from specialists and subcontractors. Taking their ideas on board required the core design team to work flexibly throughout the life of the project. Partnership for Learning Director, Roger Burton explained: 'The main challenge for the team was to allow the design to remain flexible enough, late enough in the process for the contractors' ideas to have real value.'

This project showed that integrating the supply chain into the project team increases the chances of producing a design that meets the needs of all the organizations involved. The benefits of integrating the team included:

- Better value for money – this resulted from treating quality rather than cost as the primary driver, and led to a more appropriate use of materials and systems.
- Better cost control – this came from integrating subcontractors in the design process and open book contracts so that greater cost predictability was achieved.
- Waste minimization – the integrated team's continuous involvement with the specialists reduced waste in several ways. Design time was reduced. Components and elements fitted together correctly so construction was efficient. Material waste during construction was substantially lower than normal.

Partnering reduces contractual risks

Case Study Reference: 010

A high-risk £12m project for the construction of the Tunstall Western Bypass was completed ten weeks ahead of programme, within budget and to the agreed high quality, thanks to the development of a partnering approach between client, consultant and contractor.

The challenge was to create an environment in which a combined team of staff from client, consultant and contractor worked together on and off site to anticipate and resolve problems quickly and effectively. Actions taken at the start of the project included:

- The client adopted an approach of open communication so tenderers were well informed and the brief was clear.
- When the contract was awarded, a two-day team-building workshop was held for the entire project team. The workshop focused on changing old adversarial working practices and introduced a new approach of openness and cooperation.

The key benefits were:

- The project was completed under budget and ahead of schedule.
- Claims potentially as high as £6m if a traditional approach had been used were prevented by risk management and joint problem-solving.
- The final cost was reduced by £800,000 through joint value management and value engineering.
- The client's budgetary control and contractor's cash flow were improved by dispute-avoidance procedures.
- All the concerns of local traders and residents were resolved quickly. Final accounts were agreed within a few weeks of completing the construction work.

The Pavement Team and Measuring the benefits of partnering

Case Study References: M4i 64 and 134

BAA's runway and apron construction is undertaken by a partnering arrangement between BAA and AMEC and three key suppliers. It is run by a fully integrated team that forms what is in effect a virtual company. Staff are seconded from the individual companies and share office facilities where IT and administration are provided by AMEC. 'An outsider would find it difficult to match individuals to parent companies,' remarked BAA general manager Richard Jeffcoate: 'We pick the best person for the job, irrespective of whether they're client or contractor staff.'

AMEC's Andrew Ellis explained: 'We've lived together for nearly five years now and learned how to make partnering work. The results are impressive.' BAA general manager Richard Jeffcoate added: 'It is all about getting best value for money. We moved towards partnering and an integrated team when we realized that traditional procurement methods offer very little scope to improve value for money.'

The key benefits demonstrated by careful measurements against agreed benchmarks are:

- The client is confident that the team provides value for money.
- Benchmarks were established that enable the team to measure its own performance against similar projects in the UK and overseas.
- Measuring allows the team to identify areas for improvement more easily.
- The contractor has greater predictability of workload because the client has committed to over 50 projects in five years with an average value of £2.5m.
- Cost and programme predictability improved so that projects routinely finish on time and within budget.
- Construction costs have been reduced by close to 30%.
- The total time for projects has been reduced by 30%.
- Safety performance improved and is well above industry average.
- Staff productivity increased to around 250% of the industry average.

Partnering for social housing refurbishment

Case Study Reference: Housing Forum HF175

The London Borough of Camden used partnering with its design build contracts for the refurbishment of over 2,500 properties. They formed partnering teams with two contractors, Willmott Dixon and Llewellyn.

The key benefits from using partnering in this way included the following:

- The client saved over £500,000 from a budget of £7.8m.
- In addition the fees paid by the client reduced from 14.1%, which is normal on traditionally procured contracts, to between 10.5% and 6.5% on the partnering contracts.
- Greater predictability of time: 74% of projects started on time and 70% finished on time or early.
- Far fewer complaints from tenants about the construction work. Less than 1% of tenants made formal complaints and the Council even received seven letters of commendation from tenants.
- There were no formal disputes or claims. Unavoidable extensions of time and the costs of necessary additional work were agreed quickly in the spirit of partnering.

2

Selecting firms for partnering

2.1 Introduction

Consultants, contractors and specialists should be selected for project partnering by the client's internal team. They should use best-practice selection processes and carefully thought-out selection criteria.

This chapter provides advice on selecting firms for construction projects that use partnering. In many cases, clients take the initiative in establishing project partnering and this chapter describes that common situation. However, there are cases where consultants or contractors make the first move. This should be encouraged and the guidance in this chapter applies equally to clients, consultants, contractors and specialists.

Clients experienced in partnering tend to work with consultants, contractors and specialists they know and trust. However, at some earlier point they had to select firms carefully and this chapter begins with guidance on these initial decisions. This leads into a description of the use of framework arrangements and other features of more developed partnering.

This chapter is included early in this code of practice to emphasize that in making partnering work it is important to choose the right partners. The first step in this is essentially a process of self-selection. Clients, consultants, contractors and specialists contemplating getting involved in partnering should first check that their own firm meets the selection criteria they subsequently apply to potential partners. This applies particularly to the client's internal team, which should have a confident partnering ethos.

The second reason for including this chapter early in the code of practice is that, depending on the nature of the project and the client's objectives, consultants, contractors and specialists need to be selected at various stages of projects. The individual stages are described in subsequent chapters of this code of practice. However, the same principles of best-practice selection should be used at all stages and so it is convenient to describe them in a separate chapter.

Firms considering the use of partnering should clearly accept the principle aim of partnering. This means they can answer a confident 'yes' to the following questions:

- Do you accept that working in cooperation with the firms that form a project team can provide more benefits for you than if everyone concentrates narrowly on looking after their own interests?
- Do you want the firms you work with to make a fair return for their involvement in the project?

It is prudent to check that the answers are supported by the firm's internal policies and actions on previous projects. Having confirmed in this way that they are able to use partnering, firms should chose their partners on the basis of their performance on projects similar to the one being considered. This means the selection processes should take account of their technical competence, their experience of the specific role they are required to undertake and their partnering attitudes and skills.

These criteria should be expressed in clearly defined minimum standards that the client and all the consultants, contractors and specialists involved must achieve. The selection criteria should be ambitious in including targets for better than normal performance. The next three sections of this chapter provide advice on these issues. Further guidance on the characteristics of competent teams is given in Chapter 7, checklist 20, and of effective links between work teams in checklist 21.

The specific selection criteria used on individual projects should be established by the client's internal team together with any key members of the project team that have already been appointed. This is all described in Chapter 3, which provides advice on the composition of the client's internal team.

In addition to defining the selection criteria, the internal team needs to develop selection processes that address the following issues:

- The number and type of consultants, contractors and specialists needed for the partnering arrangement to achieve the client's objectives. This depends on the nature of the project and the availability of suitable candidates.
- Ensuring that there are no major weaknesses in the partnering firms and that they provide complementary skills and knowledge.

Internal teams may be told that certain legislation or official policies inhibit the use of partnering. Experience shows conclusively that no such restrictions exist in legislation or official policies. Partnering is being used successfully in all sectors of the construction industry by all types of clients. Government at all levels is concerned to obtain best value for public money and accepts that partnering has a key role to play in construction projects. Therefore internal teams should not be deflected by suggestions that partnering is illegal or against public policy. Any such suggestions should be checked but this code of practice takes the view that partnering can be used, it should be used and it can provide substantial benefits for everyone involved.

2.2 Technical Competence

The selection processes need to ensure that selected consultants, contractors and specialists will provide technically competent work teams. The processes need to recognize that work teams good at producing well-established answers are different from those skilled at producing original answers. Selection should be based on firms' performance in delivering quality and completing projects on time.

The selection processes need to ensure that the selected consultants, contractors and specialists are able to provide the key products and services required for the project. The most important factor in this is experience in successfully carrying out similar projects to the one the client wants. The performance criteria should check the skill levels of the firm's work teams and the quality of support they get. It is important to ensure that the technical answers the firms will provide are consistent with the client's operating and maintenance policies for their constructed facilities.

The selection criteria should take account of the essential character of the project. Where an established or standardized solution is appropriate, firms selected should have a track record in delivering the answer successfully. They should have a long list of satisfied customers. They should provide detailed information about exactly what the client will get. They should be able to arrange visits to similar buildings or infrastructure including discussions with clients, local community leaders, neighbours and others likely to have been influenced by these previous projects. They should provide convincing evidence that they deliver on their promises about quality, time and cost. They should describe efficient control systems and show how they are supported by well-developed procedures. The firm should give an overall impression of solid competence and reliable efficiency.

For projects that pose unusual challenges, the firms selected need to be skilled at producing original answers to job-specific problems. Their work teams should be creative. They should demonstrate that they respond creatively to challenges and opportunities by providing examples of their own innovative designs. They should have clear evidence that they can work cooperatively with other project team members. They should have a track record of finding new answers that delight their clients and of meeting agreed deadlines and budgets.

Many of today's construction projects involve working in situations where the general public, staff, customers or others must have access. If this is the case, firms' previous successful experience of dealing with such projects should be an essential aspect of the selection processes. This is an area that produces significant risks for the uninitiated.

These criteria apply at the level of the firm but in many ways it is more important to ensure that they apply equally to work teams and individuals. This means the people who will form the project team are experienced professionals used to working together. They have the knowledge and skills needed to ensure their own firm's interests are fully taken into account in reaching project decisions. It means they have the authority to commit their firm to actions without needing to refer back to senior managers.

This last requirement may conflict with a firm's well-established procedures. Some firms protect their reputations by having decisions reviewed by a panel of senior and experienced designers. This safeguard may provide an important part of ensuring the project meets all its objectives. Provided any such arrangements are discussed during the selection processes, they can be built into the project team's agreed ways of working.

Partnering requires the levels of technical competence described above but the performance criteria should go further. Partnering often requires a change to a firm's working ethos to achieve its main purpose of delivering performance improvements. This requires firms to encourage flexible attitudes throughout the workforce and should be evident in the way they welcome and enjoy change. Ideally they should be involved in research, development and innovation.

2.3 Project Organization

The selection processes must take account of the demands made on work teams by the specific form of project organization. The client needs to understand the time, resource and risk implications of the form of project organization.

Construction uses various forms of project organization that are commonly called design build, prime contracting, general contracting, management contracting or construction management. These basic procurement options are overlaid in modern construction by the distinctive demands of the private finance initiative, public–private partnerships, design, build and operate, and other arrangements that give construction firms a long-term interest in the operation of constructed facilities. The choice of overall project organization influences what the selected firms are required to do. This means the client's internal team has to consider the specific roles of all the designers, managers, manufacturers and specialists involved in the project. Chapter 7, checklist 3 describes the main features of each approach.

In establishing selection criteria, it should be kept in mind that the client's objectives interact with the requirements of the project organization to determine the particular culture that needs to be established and fostered throughout the project. This may mean concentrating on streamlined efficiency, creative new ideas, reliable delivery, high quality, high efficiency, fast performance or some other emphasis. It is important to select firms that have the required culture so the project team can work together in a compatible and cooperative manner.

In ensuring the project organization is properly reflected in the selection criteria, the client's internal team should check that the client fully understands the time, resource and risk implications of the chosen approach. As Section 3.8 explains, the early key decisions establish the project's feasibility, which should be explicitly checked before the selection process begins. This is particularly important for clients with little experience of working with the construction industry.

The overall aim of the selection criteria should be to set up a project team able to meet the client's objectives in a manner that leaves everyone involved satisfied with the outcomes. Chapter 7, checklist 33 provides a glossary that describes the various kinds of teams used in modern construction.

2.4 Cooperative Teamwork

The selection processes need to ensure that work teams that form the project team are competent at cooperative teamwork and that key individuals have the right interpersonal skills.

The selection criteria should ensure that the firms chosen will work in cooperation with the client's own firm and those that provide work teams for the project team. The criteria should aim to identify firms, work teams and individuals clearly able to agree mutual objectives, decision-making and problem-resolution systems, and specific improvements to their normal performance. As explained in Chapter 3 these decisions are often recorded in what is called a Partnering Charter. It is therefore sensible to look at Partnering Charters used on projects undertaken by the firms being considered for selection.

The selection processes should check how firms are organized internally to support the use of partnering. Their work teams should be experienced at cooperative teamwork. It is a good sign if the firm has a senior manager acting as its internal Partnering Champion encouraging and supporting work teams. As a result firms should have a good track record of steadily improved performance on partnering projects.

Potential partners should recognize that short-term views are not compatible with partnering and that time and resources need to be invested in building up long-term benefits. Costs initially will increase but the long-term costs should be dramatically lower. This pattern of initial costs and subsequent benefits can be achieved on individual projects. The net benefits are much greater on a series of projects but they can be substantial on a one-off project.

These facts have important implications. They require potential partners to have a good performance record, the potential to change and develop, and an understanding of and commitment to partnering.

All members of the project team must be ready to make partnering work. This needs very careful consideration if a firm being considered normally adopts non-cooperative attitudes and adversarial ways. It may be that these non-partnering behaviours have been forced on them by commercial pressures and low profit margins and they genuinely want to use partnering. In such situations the client's internal team needs to consider whether the firm is ready to work in new ways and make the necessary changes in a short time. These kinds of issues make tough demands on selection criteria.

It is most sensible for clients with little or no experience of using the construction industry to use partnering only if they can employ construction firms already well experienced in cooperative working.

In addition to evaluating the firms, it is vital for the client's internal team to keep in mind that choosing the wrong individual for a crucial role could derail the whole arrangement. For each role they should consider the need for interpersonal skills, which may be even more important than technical knowledge and experience.

The following attributes are usually found in people able to use partnering successfully:

- Approachable and confident team player.
- High level of integrity and sincerity.
- Self-motivated and self-disciplined.
- Willing and able to contribute to the overall project.
- Shows commitment and enthusiasm for working openly.
- Already applying collaborative principles to existing activities.
- Willing to adapt to changing circumstances.
- Enjoys and responds positively to being challenged.
- Has the courage to do things differently.
- Enjoys being creative.
- Has the courage and honesty to state the facts.
- Can empower others to take responsibility and make decisions.
- Will support and challenge others to develop and make choices.
- Prepared to adapt behaviour to benefit the project and team.

The overall aim should be to set up an integrated project team that gives all parties the opportunity to contribute their best work. Everyone should be fully committed to cooperative teamworking. Everyone should be explicitly empowered to contribute to project decisions and assist other members of the project team. They must be willing for any individual company policies and procedures that may hinder the project's progress or success to be discussed and a joint decision made about how they will be applied to the project. An important indicator that a cooperative team has been established is that everyone is enthusiastic about being part of a team with the same overall goals and objectives.

2.5 Balancing Quality and Price

The selection criteria dealing with technical competence, project organization and cooperative teamwork should be balanced against the price.

Having determined the performance criteria that selected firms need to satisfy, the client's internal team needs to decide how they should be balanced against the price. In making this judgment the total life-cycle costs and environmental impacts should be fully taken into account. The team should also keep in mind that it is a false economy to compromise on performance or attitudes in order to get a low initial price.

Partnering can and does deliver low prices by employing competent work teams, ensuring their financial position is secure and enabling them to work to tough objectives as a cooperative team. Low prices that represent good value are not achieved by ruthless competitive tendering backed up by tough contracts. Time and time again selecting firms offering the lowest initial price has turned out to be a costly way of buying a building or infrastructure. A much more balanced approach is needed to give clients best value.

The client's internal team nevertheless has a key choice to make. This is whether to establish a fixed price or a fixed budget that they want the project team to work within.

A fixed price is the right approach when the constructed facility can be fully defined. This means that a full and clear statement of the client's needs can be produced that will not be altered during the project. It means that the site and ground conditions are fully surveyed and understood by the construction firms employed to produce the required building or infrastructure. It means there are no major non-construction risks likely to influence the project. In these ideal circumstances, it makes sense to agree a fixed price. Where a fixed price is appropriate, best practice tends to give equal weight to performance and price in evaluating potential partners.

Where the project team will develop the design of the required building or infrastructure during the project, it is best practice to agree a fixed budget based on the client's business case. The client's internal team needs to check the feasibility of this as described in Chapter 3. Given that the project is feasible, the selection processes need to establish the way the actual costs will be calculated as the design is produced. They also need to ensure that the chosen firms have effective cost control systems in place and a good track record of completing projects within budget. It is usual to allocate a weighting of 70 to 80% to performance criteria and 20 to 30% to price criteria. Chapter 7, checklist 6 provides an example of a balanced evaluation sheet for a project requiring an innovative design.

2.6 Selection Processes

Selection processes normally use questionnaires, interviews and negotiation. These are used to establish firms' track records in completing projects successfully and their established or potential abilities at partnering.

The selection and appointment of construction firms is one of the most important steps the client's internal team takes to ensure a project's success. Partnering provides clients with their best chance of getting excellent value for money and the lowest sensible final price. These are unlikely to be achieved by traditional selection processes based on competitive tenders designed to find the lowest bid price.

Once selection criteria that balance performance and price are agreed, the client's internal team needs to design the selection processes. Partnering projects normally use questionnaires, interviews and negotiations. It may require help from external consultants to design selection processes that are new to the client's internal team. Advice may come from independent experts on partnering or clients already using partnering for similar projects.

European Union procurement rules may influence some aspects of the selection processes. Most firms subject to these rules are aware of the implications but new clients should check whether they are affected and if so seek legal advice before finalizing their processes.

Selection processes should take into account the amount of work involved in the project, the nature of the necessary technologies and the possibility of future projects. A small, one-off project using traditional construction technologies can use relatively simple processes. Large, complex, difficult projects facing considerable uncertainties and requiring capital intensive prefabrication and other sophisticated technologies need more formal and thorough selection processes.

The first stage in the processes is to identify suitable firms and provide them with information about the project. Having identified firms that appear to be suitably qualified and enthusiastic about the project they should be asked to complete a questionnaire based on the selection criteria. The answers should be evaluated objectively to identify two or three suitable firms. They should be invited to a formal interview carried out by the client's internal team and any key members of the project team already appointed. The interviews should be evaluated systematically with the aim of making an objective decision about the firm that provides the best overall value for money.

The selected firm should be invited to negotiate the terms on which they will be employed. If the negotiation achieves a mutually satisfactory outcome, the firm should be appointed. If this is not possible, the client's internal team must decide what to do next. This may be to invite the second best firm for interview or to go back to an earlier stage and repeat the process from that point.

At each stage a written account of the selection process should be kept on file to provide transparency for auditors. Also detailed feedback should be offered to unsuccessful firms.

The stages in selection processes suitable for partnering are illustrated in Figure 2.1.

2.7 Identifying Suitable Firms

Suggestions about suitable firms may come from business contacts, construction industry clients currently undertaking projects locally, trade associations, professional bodies in the construction industry or other sources.

Potential partners can emerge from a variety of sources. These include the client's own personal and business contacts, and industry associations. Suggestions may come from the client's internal team, key members of the project team already appointed, construction clients undertaking projects locally, trade associations, professional bodies in the construction industry and other sources. Construction firms already using partnering may approach the client offering a better way of working. Construction journals and magazines often publish case

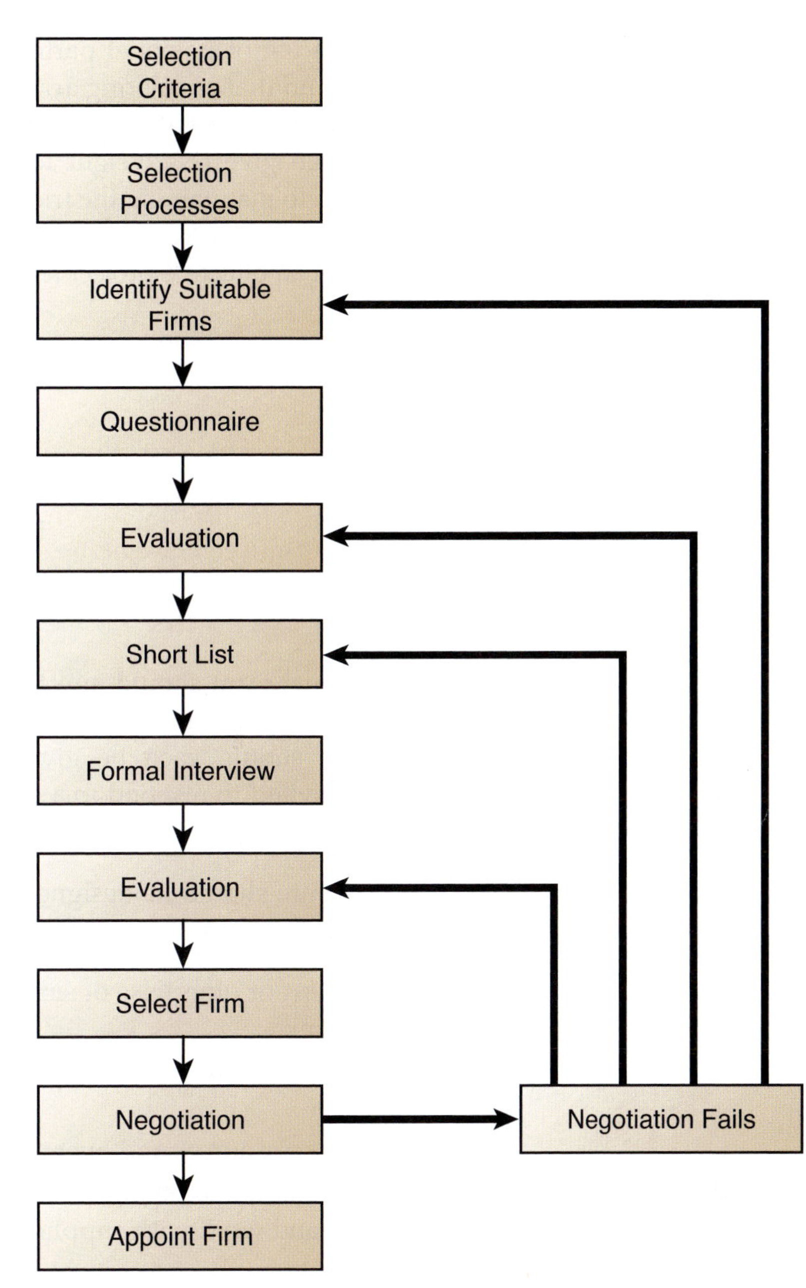

Figure 2.1 Stages in the selection process

studies of best practice by leading firms. Also, the specialized websites listed in the Bibliography help identify suitable construction firms.

A list of possible candidates should be drawn up by the client's internal team from these various sources. Having identified apparently suitable firms, the client's internal team should invite each of them to introductory, no-obligation seminars or one-to-one meetings to discuss the project and gauge their commitment. It should be made clear that a partnering relationship is desired. Precisely what this means should be discussed, emphasizing the need for cooperative teamwork and improved performance. These discussions typically deal with the need for openness, trust, effective communication, shared information, value management, risk management, mutual benefits and a joint approach to solving problems and decision-making.

In producing a list of potential partners, the client's internal team should bear in mind that partnering arrangements developed through personal friendships often fail. This is because, faced with tough decisions, people often give less weight to the needs of the partnering arrangement than to maintaining the friendship. At all stages the selection processes should be objective and concentrate on finding competent firms to form an effective project team.

2.8 Questionnaires

Five or six firms that appear to be potential partners should be sent questionnaires inviting them to describe their understanding of the project and the work they will be required to do if appointed. They should also be given a statement of the selection criteria. The results identify firms worth interviewing.

The client's internal team should aim to draw up a list of five or six apparently suitable firms that have confirmed an interest in being appointed. They should each be given an outline description of the project and invited to respond to a questionnaire dealing with the selection criteria.

The questionnaire should be designed to obtain objective information about the firm's relevant experience, including:

- The success or otherwise of similar projects they have worked on.
- Their policies towards life-cycle costs and environmental impacts.
- Their size, resources and overall organization.
- Their technical and organizational skills including their ability to select and work with suppliers and subcontractors.
- Their health and safety record.
- Their financial stability.
- The basis on which they are prepared to be employed.
- Detailed information about how they suggest the price should be determined.
- References from clients and construction firms they have worked with on similar projects.
- How they measure the firm's performance and their current targets for improvement.
- Evidence of their ability to innovate.

The client's internal team should each evaluate the responses to the questionnaire using the agreed selection criteria. The separate evaluations should be collated and the results fed back at a team meeting where any large differences in the evaluations can be discussed and a consensus sought. It may be necessary to seek further information from some of the respondents before a final evaluation can be made. It may be necessary to hold further meetings to establish an evaluation the whole team can accept. Whatever actions are needed to arrive at a decision, the team needs to keep in mind that their aim is to identify two or three suitable firms.

2.9 Interviews

Interviews are used to select which firm should be invited to negotiate the terms of a contract.

Each firm selected on the basis of the questionnaire responses should be interviewed. This allows the client's internal team to meet people who will form part of their project team if the firm is awarded the work.

If there is any doubt about the experience or capacity of members of the client's internal team to participate in the interviews, they should be given training or carefully briefed before the interviews take place.

The firms to be interviewed should be given a clear description of how the interview will be conducted and what is expected of them. Each interview should begin with the firm's representatives making a presentation describing their understanding of the project and their role within it. They should introduce the key people who will be involved if they are selected and describe how they propose to undertake the required work and how they expect to fit into the project team.

The client's internal team should ask questions to clarify information from the questionnaire or given in the presentation. Then they should ask open-ended questions that allow firms to demonstrate their knowledge, abilities and experience.

The interview should be used to establish each firm's partnering experience and potential. This means asking questions about their policies and strategies, how they make decisions, the way they deal with problems, how they balance quality, time and cost, including life-cycle cost, and their track record in working with others in improving their joint performance. The aim is to be certain that the firm selected will make a partnering arrangement work successfully.

After the interviews each candidate should be objectively scored against each of the selection criteria to decide which of them should be invited to negotiate the terms of a contract. As with the evaluation of the questionnaires, the individual scores should be discussed until a consensus is reached. The decision and its basis should be recorded and all the firms provided with feedback on the results.

2.10 Negotiations

Negotiations should concentrate on agreeing what the firm is required to do, set up a basis for establishing fair terms and establish the groundwork for partnering.

The selected firm should be invited to negotiate the precise basis on which it will be employed. All the well-established principles of negotiation are likely to come into play and a normal approach to business negotiations should be used. It is important that this process is not seen as an opportunity to gain advantage at the expense of the other party. The aim is to establish a robust basis for partnering.

When the project is to produce a building or infrastructure for which there are well-established designs, negotiations should concentrate on defining the new facility and a fixed cost and time. These should be similar to the negotiations surrounding say the purchase of a new car. There will be a range of options and standards and widely known market prices and times. Essentially the client is deciding which of the facilities on offer best meets the business need. At the same time the construction firm is ensuring that the client fully understands the choices available and that the final choice will indeed give the client everything they expect.

When the project has to produce a new design, negotiations concentrate on ensuring that everyone involved understands the client's objectives, the resources to be provided by the construction firm, the way the project team will work together and the completion date and budget. These negotiations are likely to concentrate on issues concerning people, teams, skills, experience and attitudes. The aim is to ensure that there is a common understanding of the client's objectives and the financial and time targets so that an effective new member is brought into the project team.

In all cases it is important to review the major sources of risks to ensure that everyone understands how each kind of risk is likely to be managed. This usually means that all the information that exists about the site, ground conditions and main services is fully considered. Similarly if the project involves working in an existing building or altering existing infrastructure, all the implications must be discussed until there is a common understanding of how problems and risks will be managed.

The outcome of the negotiations is usually a decision to enter into a formal contract. Advice on the various standard forms of contract suitable for partnering projects is given later in this chapter. It is important not to let detailed, nit-picking discussions of contract clauses distract members of the project team from their proper work of producing the required building or infrastructure. The Egan report's recommendation that the construction industry should learn to work without using formal contracts was included with the aim of avoiding the waste involved in spending substantial amounts of time and resources in tough contract negotiations. Traditionally these negotiations are very expensive not just in the direct costs but in the even higher costs of damaged motivation as members of the project team fight for some narrow advantage under the misguidance of lawyers and bureaucrats. Such negotiations are incompatible with partnering and it is important to ensure that the negotiations are not allowed to degenerate into such wasteful battles. Indeed if this happens, the client should carefully consider whether they have selected the right firm.

2.11 Competitive Tenders

Selection processes can include competitive tenders if the client wants this. The competition should be part of selection processes that balance firms' performance and partnering achievements or potential with the price.

Some clients have to use competitive tenders to select the firms they award contracts to because of organizational policies or legal rules. Others feel uncomfortable if they have not tested the market. Whatever the reasons, competitive tenders can be used with partnering. They provide no advantage and can easily create potential problems for the project team. However, some clients decide to use competitive tenders.

The construction industry has well-developed advice about best practice competitive tendering. Material on this subject published by the Construction Industry Board and listed in the Bibliography deals with the selection of consultants, contractors and specialists including subcontractors and is sound. The advice includes two-stage tender assessment processes that separate performance and price criteria. This helps the client's internal team concentrate on selecting firms that will provide effective members of the project team. Various public bodies have other well-developed procedures and they should be used straightforwardly. Essentially they replace the questionnaire, interview and negotiation stages of the selection processes described in this chapter. Chapter 7, checklist 7 provides a list of issues to be kept in mind when using competitive tenders on partnering projects.

Bid documents should describe the client's objectives and the proposed partnering arrangement, its objectives, scope and the length of time it is intended to remain in force. It should explain how bids will be evaluated. It should ask bidders:

- Why do they want to use partnering?
- What is their experience of partnering?
- How will they select staff to work on the project?
- What is their main contribution to the partnering arrangement likely to be?
- What actions will they take to put partnering into effect?
- How do they expect problems to be resolved?
- How do they measure key performance indicators?
- How will they make performance improvements?
- How will they suggest project costs should be calculated?
- How does the contract relate to partnering?

These matters may need to be clarified at an interview or some other face-to-face meeting with the people who will undertake the firm's role in the project. It is vitally important to ensure in these discussions that the lowest bidder has not submitted an unrealistically low price with the intention of recovering the shortfall during the project.

All the firms should be told the tender result quickly. As soon as a contract is agreed with the selected firm, the unsuccessful firms should be told. They should be given detailed feedback on where they fall short of what is likely to be required on future projects, particularly if they lost out because of weaknesses in partnering experience or aptitude.

2.12 Framework Arrangements

Framework arrangements provide choice in appointing firms for projects using partnering. At the same time they encourage construction firms to invest in improving their performance by providing a realistic chance of a regular and stable workload.

Framework arrangements are used by some major clients to provide a pool of competent consultants, contractors and specialists for their construction projects. These arrangements establish a basis for negotiations over future contracts with a limited number of firms. Essentially they provide standing offers by firms to provide specific goods or services on predetermined terms and conditions which remain valid during the lifetime of the agreements.

Framework arrangements take various forms. They need not be binding contractual agreements and they do not necessarily imply a promise of work in the future. In such cases, although the agreements involve no commitment to purchase, they commonly specify the terms and conditions of the eventual contract that will apply when goods or services are purchased. These arrangements usually run over a specified period of time. No overall contract is formed because no consideration is given as part of the framework arrangement. Suppliers can terminate the standing offer with immediate effect at any time by giving notice to that effect to the buying firm. Obviously they have to fulfill any contract for an individual project formed before termination.

Other framework arrangements include a contractual commitment to purchase a particular minimum volume or value of goods or services during the period of the framework. Sometimes the formal contract is based on a consideration of a purely nominal sum paid by the buying firm to the framework supplier. These arrangements run over a specified period of time and are contractually binding.

In all these various arrangements, a separate contract is formed each time the agreement is used.

Framework arrangements are established following a normal tender exercise in which suppliers are invited to bid to supply a stated quantity of goods or services. Following the tender exercise the buying firm forms a framework agreement with between one and four suppliers. Decisions about how many firms to include should take account of the type of construction work, the firms' capacity and the local construction market.

The award of subsequent contracts under the terms of the framework agreement are straightforward if only one supplier can meet the specific requirement or where one supplier clearly offers the best value for money. In these circumstances a contract is negotiated and formed with that supplier. Where two or more of the suppliers can meet the particular need, a mini selection process based on the principles described in this chapter should be held with those suppliers. The basic terms should not be renegotiated, nor should specific requirements depart significantly from the general terms of the framework agreement. This is particularly important if European Union procedures apply because any significant departure from the terms of the agreement could be used to force the buyer into a fresh tendering exercise.

Framework arrangements have a number of advantages, including:

- Establishing a basis for long-term relationships which help establish partnering arrangements.
- Allowing specialist buyers to negotiate the best value for money for goods and services that are used on a number of projects which can then be purchased as and when required.
- Allowing large firms to utilize their purchasing power to get competitive prices.
- Making it easier to ensure that procurement adheres to official purchasing policy and standards and complies with appropriate legislation.
- Reducing the need to conduct individual tendering exercises.
- Reducing the procurement time on individual projects.
- Providing an assurance of consistent quality and standards.
- Making assured and early delivery more likely.
- Ensuring that major problems are tackled quickly, if necessary at a senior level.
- Making a reliable after-sales service more likely.
- Providing a basis for steady continuous improvements in performance.
- Allowing life-cycle costs to be considered carefully.
- Encouraging sound environmental impact policies to be developed.

2.13 Supply Chains

Selecting firms for project partnering should take account of the competence of their supply chains and the extent to which they make use of partnering.

An important characteristic of firms experienced in partnering is that they develop efficient and cooperative relationships with their main suppliers. Leading firms in the construction industry have adopted supply chain management and their key suppliers contribute to project decisions as full members of a partnering team. Client's internal teams should look for evidence that the firms they select have well-developed supply chains. This should be one of the most important selection criteria particularly where framework arrangements are used.

Well-established supply chains have robust processes aimed at improving the efficiency of the supply chain. These should cover every aspect of the supply chain processes including procurement, design, manufacturing and installation. They should aim to streamline each component of the supply chain and improve all aspects of quality.

When a project uses a well-developed design, there should be well-developed supply chains in place for all the major elements and systems. The lead firms in each supply chain should be appointed early and fully involved in the partnering arrangement.

The situation is more complicated when the project requires an individual design. It is always sensible for the members of the project's core team to be appointed early so that they can agree with the client's internal team the supply chains most likely to be needed. Suitable firms can be appointed on the basis of flexible contracts that can be terminated if it becomes clear that the decisions about supply chains have been invalidated by subsequent design decisions. The lead firms in each supply chain should be fully involved in the partnering arrangements. There are many advantages in this approach, not least in ensuring that project costs and risks can be identified early. It allows the whole team including those responsible for the design and construction processes to be integrated from the outset. These are substantial benefits that help ensure the successful completion of projects even if some of the firms have to be changed because the design develops in unexpected ways.

The early appointment of the lead firms in key supply chains is particularly important where the project work includes altering an existing building or infrastructure. Direct communications between representatives of local people and those undertaking construction projects can deliver dramatic improvements in the overall level of satisfaction. Experience shows that early appointments allow local people and organizations to discuss issues of access and safety. The discussions are particularly beneficial in determining how work can be done in or around occupied premises.

The involvement of construction firms from an early stage gives clients time to encourage them to invest in local employment, training and development. Many major clients and construction firms see it as important that they foster local communities in this way.

The maximum benefits and efficiency come from integrated project teams comprising fully integrated supply chains selected to meet the specific requirements of the project. This should be the aim of the selection processes whenever such supply chains exist. However, the dynamics and diversity of the construction market, and its tendency to use individual designs even when perfectly good answers already exist, inhibit the natural emergence of fully integrated supply chains. An increasing use of partnering has made the emergence of permanent supply chains more likely and this needs to be encouraged in the interests of the industry and its clients.

In the meantime, the answer for many projects lies in recognizing the existence of supply chain modules and mix-and-match mini supply chains. These allow project teams to be assembled from a matrix of modules and mini supply chains to meet specific project needs. In many cases lead firms in sectors of the construction industry structured in this flexible way work through what are in effect flexible long-term partnering arrangements. Selection processes need to take account of these developments and encourage the lead firms to search for innovative ways of improving the industry's performance.

2.14 Contracts

Contracts should support partnering. Many standard forms in common use fall short of what is required. The contract should be chosen taking account of the practical implications, particularly the financial implications and the impact on work teams' ability to partner confidently.

Many firms want formal contracts in place for construction work.

Partnering agreements usually create a partnering charter that sets out the principles, attitudes and ideals that will characterize the arrangement. This is normally produced at the first partnering workshop as described in Chapter 3.

Many of the basic requirements for partnering to be successful are not dealt with in standard forms of construction contracts. The attitudes and patterns of behaviour that develop as people, teams and firms work together cannot be predetermined in formal legal terms. They have to be deliberately fostered, monitored and encouraged. Partnering is based on the idea that this effort is worthwhile because people working as a cooperative team achieve far more than those working in traditional arrangements based on individual rights and responsibilities defined in contracts. Negotiating the terms of a formal contract tends to destroy partnering attitudes. Working to rules and procedures defined in a standard form of contract inhibits partnering behaviour.

Nevertheless many firms think they need the protection of a formal contract. As a result a number of unsatisfactory arrangements are commonly used. Without doubt the worst approach is to use a traditional standard form of contract but to have an implicit agreement that its terms and conditions will be ignored in using partnering. The temptation to fall back on contract provisions when a serious problem arises instead of working as a team to find the best answer is too much for many people. Many potentially good partnering arrangements are inhibited or destroyed in this way.

When some of the firms involved in a partnering arrangement insist on a formal contract, it is important to choose a form that is reasonably consistent with partnering. The contract should deal with the following key issues:

- Customer satisfaction
- Construction firms profits

- Early appointment of all key members of the project team
- Core team working arrangements
- Quality, time and cost control systems
- Open book accounting
- Joint value and risk management
- Problem resolution mechanisms.

One of the key benefits of effective partnering is being able to raise issues or areas of concern at an early stage when they can be resolved by discussion and real agreement. This is what the first partnering workshop does in producing the partnering charter.

The partnering charter describes the essential agreement between the partnering firms and should be included in the formal contracts.

It is reasonably well established that the Project Partnering Contract (PPC 2000), its derivatives, and the NEC Engineering and Construction Contract with its Partnering Option are the most appropriate for partnering arrangements. Both are supported by good advice listed in the Bibliography, and there is sufficient experience of using them for clients and construction firms to be confident that they will not inhibit the use of partnering.

Traditional contracts like the Joint Contracts Tribunal (JCT) Standard Forms of Contract are often used with a Partnering Charter attached. JCT has not produced any amendments to accommodate partnering. They have issued a practice note which advises that inserting a binding partnering agreement into existing forms of contract is inappropriate and could lead to potential problems. The practice note recommends that parties using partnering should enter into a non-binding partnering charter and provides an example of a JCT Non-Binding Partnering Charter. This creates a far-from-ideal situation that inevitably includes some serious ambiguities but some clients have to work with traditional contracts because they are the only ones accepted in their organizations. They should work to change their firm's policy as quickly as possible. The long-term aim should be to get authority to work without formal contract terms and conditions on the basis of carefully agreed partnering charters and the common law. If that cannot be agreed, they should at least persuade their firm to use PPC 2000 or the NEC Partnering Option.

Changing outdated policies in this way is very important in ensuring that partnering succeeds. Senior managers should recognize that barriers to change can be overcome and the following actions should help:

- Seek help and guidance from people experienced in using partnering on similar projects.
- Consult widely on changes to existing procurement and contracting processes.
- Recognize that doing things as they were done in the past is habitual and changes need to be planned and managed at a senior level.

The reason for the tough approach described in this section is that anyone using a form of contract that includes terms and conditions that inhibits or neutralizes the partnering approach should not be surprised if partners resort to adversarial behaviour when problems arise. This is particularly likely if contracts are awarded on the basis of unrealistically low bid prices. This creates a temptation to exploit problems in the belief that a bad financial position can be improved by making claims. It is the nature of construction that problems will arise. Partnering provides the best approach currently available to handling all the difficult issues that construction projects throw up and it is foolish to inhibit this most effective approach by using an out-dated contract.

2.15 Project Insurance

Project partnering is helped by project insurance that allows firms and work teams to concentrate on doing their best work.

Project insurance can give firms and work teams the confidence to realize that they do not need copious records to defend their own decisions and actions or establish the basis for claims against the client or other members of the project team. In this way it provides an important part of best practice project partnering by allowing everyone involved to concentrate on doing their best work in the interests of the project.

Project insurance helps deal with a number of problems. Amongst these is the problem that the benefits of partnering are not widely recognized by the professional indemnity and latent defects insurance markets. Most underwriters seem unwilling to believe that partnering can actually reduce risks and indeed are more concerned that it blurs the edges of responsibility. The whole legal framework surrounding construction projects requires professional consultants to offer collateral warranties to financial institutions, prospective purchasers and major tenants. There is little sign of them being abandoned in order to facilitate partnering. The situation is more encouraging in the latent defects market, which is beginning to accept that cooperation between consultants, contractors and specialists can reduce risks. They still insist on being informed about design innovation and technological change and this can restrict what project teams are able to deliver through cooperative teamworking.

The best way of avoiding these breaks on effective partnering is for the client to take out insurance for the project as a whole in place of the individual policies taken out by all the individual firms involved. Often the key is persuading insurance companies that partnering does not add new liabilities and may indeed reduce risks. This needs discussion and negotiation but it can be done and it helps provide a sound basis for partnering to succeed.

2.16 Stakeholders' Approval

The client's internal team should ensure that the client, financiers and other stakeholders have opportunities to influence the selection processes and confirm their support for the firms selected and the basis on which they are employed.

Partnering is based on open communication so that problems and risks are identified early while the project team has time to find the best answers. This is vital throughout the project team, including the client's internal team. It is equally important to include all the stakeholders likely to be influenced in any way by the proposed construction project. This open communication extends to the selection of the firms that will form the project team.

The stakeholders do not all need to be directly involved in the selection processes. It is for the client's internal team helped by key members of the project team to put the selection processes into effect. However, all the stakeholders should be told which firms are being considered at each stage in case any of them have important information that may influence the selection processes. It is very unhelpful in any project to appoint a firm only to discover they are locked in a messy dispute with the leaseholder, an influential neighbour, the local authority or even worse, one of the project's financiers.

The various interests inside the client's organization should also be kept in the picture. It is important that staff at all levels, including senior management, know what is happening. Customers, suppliers and tenants may have real concerns that should be dealt with head on. Social housing authorities have led the way in undertaking wide consultation before construction firms are appointed to ensure that selected firms will work with tenants and residents' representatives to deal with such issues as access for elderly people and those with special needs. They use public meetings, visits to individual houses, open days and dedicated phone lines to help construction firms get a clear focus on customer service issues. The aim should be to ensure that the plans for construction are widely understood and as far as possible seen to be positively beneficial.

Everyone with a legitimate interest should be told that the project is being considered and given access to clear information about the project as each main stage is reached. There are many effective ways of doing this. Basic information can be made available in a nearby building open to the public, a website can be set up, a manager can be appointed to ensure that stakeholders are informed or a site office can be used to provide information. The aim is to ensure that all the stakeholders have access to information about firms being considered as members of the project team. Then provision has to be made to deal with comments and criticisms. These should be taken seriously and every attempt made to ensure that each point is understood, properly considered, and the stakeholder concerned knows this is happening and is given honest and clear feedback.

These actions all help to ensure that the selection processes achieve their overall aim of establishing a project team that has the full confidence of everyone with a serious interest in the project.

Repeat business nurtures continuous improvement

Case Study Reference: 254

Office park developer MEPC spent £11m on a series of eight office buildings at Basingstoke's Chineham Business Park over five years. This programme was used to refine its office 'product' using a partnering arrangement with contractor Balfour Beatty and mechanical and electrical contractor Crown House.

'Company boundaries do not restrict partnering with trade contractors,' says Balfour Beatty Project Director Nick French. 'It depends on what they're good at and their track record with us.'

The partnering arrangement stemmed from *Hazlewood*, a traditionally tendered job started in 1997 based on the client's design, with Crown House as the mechanical and electrical subcontractor to main contractor Balfour Beatty. *Rosewood* quickly followed, but this time MEPC negotiated a target cost. The team hit all MEPC's time, cost and quality targets.

In 1999, MEPC came directly to Balfour Beatty to negotiate the next phases, *Ashwood* and *Maplewood*. Balfour Beatty automatically invited Crown House to work on the bid. Crown House's project director, Viv Blandford, recalls: 'MEPC gave us a concise scope of works – just one page.'

In 2001, MEPC ordered *Redwood*, comprising two buildings that required piled foundations. The relationship with the contractors had now matured to 'partnering status', with integrated team working and a risk-sharing pain/gain agreement. Value engineering at *Redwood* produced a saving of between 3 and 4%. MEPC opted to reinvest this in a higher specification curtain wall system.

MEPC's manager at Chineham Business Park, Mark Younger, explained their approach to negotiating repeat contracts: 'It's about streamlining and refining the process so that we are confident of getting what we want.'

This case study showed that keeping winning teams together results in marked improvements in performance. Everyone involved understood that repeat business and continuous improvement go hand in hand. The benefits included the following:

- The partnering arrangement's cost/sq. ft was about 15% lower than traditionally procured offices, and was still falling.
- The rate of design (measured in sq. ft/week) doubled and the rate of construction almost tripled, over three phases of the development.
- Delivery on time and budget was achieved consistently in the partnering arrangement.
- The partnering arrangement normally achieved nil defects at handover.
- There was a 35% rise in productivity (measured in hours/ 1000 sq. ft).
- Balfour Beatty and Crown House have both earned higher than normal profits from this partnering arrangement.

Selecting a contractor for housing innovation

Case Study Reference: Housing Forum HF141

The Amphion Consortium, consisting of 22 housing associations, used a rigorous selection process to appoint a contracting partner, Partnerships First (formerly known as Beazer Partnerships). Their agreed objective was to make radical improvements in housing design and building systems for the social housing sector.

Amphion's invitation to tender was prepared by a representative sub-group of the consortium and its consultants. The aim of the document, which was 23 pages long, was to enable the client to examine in detail the bidders' proposed construction technologies, their sustainability and the procurement systems they used.

The invitation to tender was advertised in the building press and the client received 40 requests for documentation. Fifteen tender submissions ensued.

Six companies met the initial criteria of a clear commitment to Egan's principles, a sound approach to sustainability and sure-footed and robust financial performance.

During and after the interviews, each contender was evaluated using the value management methodology named SMART, which helped assess the company (abbreviated from the five criteria of 'specific, measurable, achievable, realistic and time-related'). Three companies were chosen.

Detailed data was collated on all three contractors. Amphion's selection criteria gave a 75% weighting to 'quality' and a 25% weighting to cost. The outcome was the appointment of Partnerships First, then known as Beazer Partnerships. The contractor partner selection process took six months in all.

In its first 15 months of operation, Amphion made significant progress towards Egan's targets of reducing capital cost and construction time by 10%, reducing defects and accidents by 20% and increasing productivity by 20%. In 2001 the partners also overcame the uncertainty provoked by Persimmon's takeover of Beazer by mounting a successful management buyout. As a result, the partnering arrangement has been extended for another year, with the potential of a further three years thereafter.

Council adopts PPC 2000 in tricky school build

Case Study Reference: 252

Brighton and Hove City Council used partnering to procure design and construction services to build above classrooms where children continued with lessons at two existing schools (Portslade Infants School and Downs Park Special School). Both schools needed extra classrooms, yet neither had land for more buildings. The only way was up. Fortunately the classrooms could be erected at first floor level without disturbing the ground floor structures or the users.

This was the first project to break through the barrier of the council's standing order blocking early contractor involvement. Nigel McCutcheon, Architecture and Design Manager recalled: 'We could not see any other practical way of building the extensions within the time and budget available. The brief was technically difficult and we really needed the contractor's advice early to ensure the design was easy to build and provided the best value for money.'

Officers sought advice from the Business Engineering Group at the University of Southampton, leading to a two-stage selection process. The first stage was an open invitation with questionnaires evaluated on a 75:25 quality:price split. For this purpose the price element was the sum of the fee to participate in the design development plus overheads and profit based on the client's estimated cost. Four bidders made it through to the second stage where each was judged on a presentation and question and answer session. At this stage, bidders were asked if they thought that the stated budget was sufficient. Llewellyn (now part of Rok Property Solutions) was selected to join the project team.

The City Council used this project to trial PPC 2000, the revolutionary form of contract for project partnering, described by Sir Michael Latham as 'the full monty of partnering and modern best practice'. The project was also the test bed for a more proactive process of dialogue with the schools in which head teachers became 'part-time members' of the project team. The benefits included the following:

- Many options were explored before the team settled on the design and agreed maximum price within the City Council's budget.
- Quick construction methods were devised to meet very tight time requirements.
- The design provides low maintenance, energy-efficient buildings.
- The project team members and the council's legal and audit teams all found PPC 2000 easy to use.
- Variations were down 75% compared with previous schools.
- There were only two defects recorded in each school at handover.
- Both school extensions were completed on time.
- The final account was settled at the agreed maximum price.
- Dialogue with head teachers led to education and construction co-existing smoothly on congested sites.
- The council is entirely satisfied with partnering and PPC 2000 and extended the original partnering arrangement at the two schools to provide a £2.0m sports hall and fitness suite at a local secondary school.

With this trial project now completed, a further five-year strategic partnership was procured using PPC 2000. The strategic five-year framework team was working on three school extensions of combined value £2.6m and further projects were in the pipeline.

Contractors collaborate with council to make homes decent

Case Study Reference: 258

The government's Decent Homes target is that by 2010 all social housing will be wind and weather tight, warm and have modern facilities. Modern central heating is essential to meet this standard. Looking for a system that would accelerate upgrades, reduce costs and boost customer satisfaction, Portsmouth City Council abandoned its traditional reliance on selecting contractors on the basis of the lowest-price and using tough, inflexible contracts. Instead they decided to use strategic partnering.

The idea was to create a framework that encouraged innovation and collaboration. Portsmouth City Council began their use of strategic partnering by forming two separate arrangements, one with United House and the other with Clenmay. Both arrangements used ECC Option C (target cost with activity schedule) with open book accounting and a pain/gain incentive to save money. The contracts were renewable annually, subject to reaching agreed key performance indicator targets, which were ratcheted up each year. The benefits included the following:

- This new way of working delivered costs for Portsmouth City Council that were lower than they were paying two years earlier.
- The number of visits to complete each job reduced from six to less than five and the call-back rate because of defects plummeted from 25% to under 5%.
- Customer satisfaction rose from 80% to more than 95%, reflecting less disruption and more opportunity for involvement.
- 5% of the site workforce were recruited locally via a project initiative to train technicians.
- About 65% of waste was recycled.
- Portsmouth City Council needed fewer staff to oversee the work.

Actions at the start of projects

3.1 Introduction

This chapter describes actions taken during the inception, feasibility and strategy stages by clients and project teams using project partnering.

Construction projects move through distinct stages. The sister publication, *Code of Practice for Project Management for Construction and Development*, provides a good description of the stages in construction projects, illustrated in Figure 3.1.

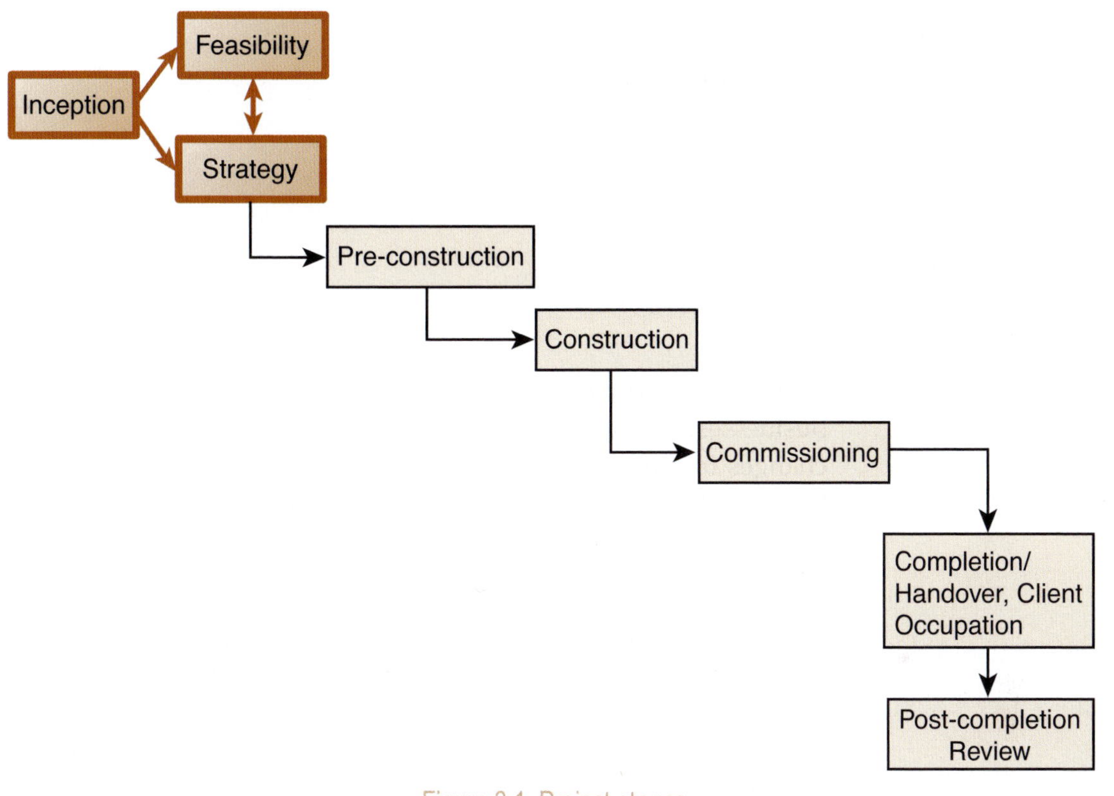

Figure 3.1 Project stages

Objectives of the Early Stages of Construction Projects

Inception	Inception begins when the business case for investing in a new building or infrastructure is accepted by the client. The objective is to turn the client's business case into a formal statement of the client's objectives for the required building or infrastructure and appoint the client's internal team.
Feasibility	Feasibility ensures that a suitable site is available, and produces a project brief, a design brief, a concept design and a funding and investment appraisal. The objectives are to determine whether it is feasible to undertake a construction project consistent with the client's objectives; and if so to produce the first version of the project execution plan.
Strategy	Strategy interacts with the feasibility stage. Its objectives are to decide the project organization, the control systems and procurement approach to be used and appoint the project team. The decisions form the basis for the first version of the project handbook.

This chapter deals with the first three stages: inception, feasibility and strategy. The objectives of these stages are listed in the box on the previous page. The objectives apply whether projects use partnering or project management but the way the objectives are achieved is different.

A key event in the early stages of projects using partnering is the first partnering workshop. This is where the project team reviews and agrees the overall project objectives and decides in detail how they will set about achieving them. The actions leading up to the first partnering workshop and at the workshop are the subject of this chapter.

3.2 Client's Internal Team

The client sets up an internal team responsible for ensuring the client's interests are taken into account throughout the project. This is led by the project sponsor who has the authority to make decisions and take actions on behalf of the client. It includes people who understand the relevant business issues and construction implications.

Construction projects arise from developments in organizations' businesses. The decision to accept a business case that may require changes to the organization's constructed facilities is normally taken by the organization's executive, which is not directly involved in project processes.

In making the initial decisions the senior managers and directors should consider the possibility of combining several projects into a programme. This can make the work more attractive to consultants, contractors and specialists because the greater volume of work reduces the costs of winning projects and understanding how the client operates. It also justifies investing in searching for the best answers and devising ways of putting them into practice as efficiently and reliably as possible. All this means that the client should get better value for money from consultants, contractors and specialists committed to satisfying a major client.

Once the initial decision is made, the organization should appoint the project sponsor. This is the individual given formal responsibility for making decisions and taking the actions needed to maintain progress on the project on behalf of the client. This requires them to have direct access to all the people in the client's organization making key decisions and the authority to take day-to-day executive action.

The *Code of Practice for Project Management for Construction and Development* recommends that a project manager is appointed at the inception stage. Partnering does not require a normal project manager's role. It is a characteristic of cooperative teamwork that the most appropriate person takes the lead depending on the stage of the project and the issues being considered. In some case a consultant project manager may be employed to be the project sponsor if no suitable person is available internally. It is important that they do not confuse this role with a normal project management role.

The project sponsor takes the lead in forming the internal team. It needs to include people who understand the relevant business issues and how they translate into physical requirements. The client's organization is likely to include people who understand the business trends and influences, industry norms and financial constraints although it sometimes makes sense to draw on external expertise. The construction advice needed by the internal team may also be provided by in-house specialists but often requires consultants to be appointed.

Construction Professionals' Roles at the Early Stages of Projects

Architects	Architects can advise on the size, shape, appearance and quality of buildings and the architectural impact on neighbouring streets and public spaces. Architects can advise on the benefits of original design and designers having a leading role in project teams.
Construction managers	Construction managers may operate as consultants or management contractors. They can advise on the quality, time and cost implications of designs and whether they can be built efficiently and safely. They can advise on the benefits of project teams including an independent management role alongside design teams creating original designs.
Design build contractors	Design build contractors can advise on the benefits of an integrated design and construction service to provide a single point of responsibility for producing a new facility that meets the client's objectives. They can advise on the benefits of straightforward buildings or infrastructure that can be delivered efficiently and quickly.
Engineers	Engineers can advise on the safety and stability of design for the structural elements and the systems that provide all the services that new buildings and infrastructure require. They can advise on the benefits of original design and designers having a leading role in project teams.
General contractors	General contractors can advise on the quality, time and cost implications of designs and whether they can be built efficiently and safely. They can advise on the importance of complete and timely design information.
Quantity surveyors	Quantity surveyors can advise on project budgets, cost planning and control systems, the production of the information needed to invite competitive bids and the various forms of standard contracts.
Project managers	Project managers can advise on efficient ways of organizing, planning and controlling all aspects of construction projects to ensure they satisfy the client's quality, time and cost requirements.
Specialist contractors	Specialist contractors can advise on the performance, quality, time and cost of their particular system or element of new buildings or infrastructure.

The choice of consultants to provide a strategic understanding of construction is determined by the nature of the required facility. The box below lists construction professionals who may make useful contributions to the early stages of projects. The internal team may decide to interview construction experts, commission specific advice or make them members of the internal team depending on how well the internal members understand the construction opportunities and issues.

Selection of members of the internal team should be as rigorous as the selection of consultants, contractors and specialists for partnering projects described in Chapter 2. It is particularly important that they have a well-developed partnering ethos and work on the basis of cooperative teamwork with each other and the project team.

3.3 Client's Objectives

The client's internal team produces a formal statement of the client's objectives.

The internal team turns the business case into a formal statement of the client's objectives for the required building or infrastructure. This should describe the building or infrastructure in the client's own terms. The questions listed in Chapter 7, checklist 1 provide guidance on the issues that should be considered. In answering the questions, the internal team may appoint consultants to provide advice on design, construction, finance, planning regulations or other matters.

The internal team should consider various ways of providing the physical facilities required by the new business need. These may include outsourcing work, reorganizing existing facilities or the way they are used, extending existing facilities, leasing or buying an existing facility and commissioning a new building or infrastructure. Each option should be considered in a process that challenges and questions its impact on the business need. The internal team should think outside the obvious and conventional answers in a constructive search for the best answer.

If the internal team decides that construction work is required they should identify the internal and external stakeholders. The internal team should determine for each stakeholder their level of commitment or opposition and establish how they want to be involved or consulted. The composition of the internal team may well need to be changed at this stage to reflect the interests of key stakeholders.

The client's internal team discusses the proposed construction work with all the key internal people. They may also set up small discussion groups comprising representatives of the main interests. The groups usually include representatives of the people who will use the building or infrastructure, finance managers and facilities managers. It may be sensible to include representatives of the customers, shareholders, unions and neighbours.

The discussion groups should be encouraged to make suggestions about the building or infrastructure. Good ideas may be generated by inviting two or three people from the construction industry to talk at the start of these meetings about interesting buildings or infrastructure. Similarly, experts can be invited to describe how buildings or infrastructure can make organizations more efficient, provide better services to customers, reduce life-cycle costs and environmental impact, enhance the local environment and help the local community. It may help to arrange visits to existing buildings or infrastructure that provide a range of facilities more or less relevant to the organization's work. The aim is to trigger ideas and build enthusiasm for the project.

Out of all these discussions, the internal team produces the statement of the client's objectives. It should identify the key objectives for the building or infrastructure in a prioritized list. The objectives are sometimes expressed as measurable value criteria to provide a basis for evaluating designs, quality standards, programmes and budgets. They establish the general nature of the project to guide subsequent decisions.

3.4 Key Choices

The client's internal team makes three key choices which affect the actions that will need to be taken by the project team as they put partnering into practice.

Once the statement of the client's objectives is formally agreed, the internal team should make the key choices described in Chapter 7, Checklists 2–4. The choices are crucial in determining the nature and quality of the building or infrastructure to be produced and the overall strategy of the project team. It is therefore essential that the key choices are considered carefully and robust decisions made, especially if the project team subsequently appointed is new to partnering. Project partnering is tough and if the project team has the wrong overall strategy, partnering may well be an early casualty of the ensuing confusion.

In making the key choices, the client's internal team establishes the project's feasibility. That is, they establish whether it is realistic to expect a building or infrastructure to be produced that fully meets the client's objectives.

The client's internal team is responsible for assembling the information needed to guide the client in making the key choices. This may require various tests and studies of the proposed site for the building or infrastructure. They may seek outline planning permission from the local authority to ensure that the client will be allowed to use the intended site. The *Code of Practice for Project Management for Construction and Development* provides detailed advice on site investigations that is summarized in the box on the next page.

Activity	Action by
Site surveys	Land surveyor and structural engineer
Geotechnical investigation	Ground investigation specialist
Drainage and utilities survey	Civil engineering consultant
Contamination survey	Environmental and/or soil specialist
Traffic study	Transport consultant
Adjacent property survey	Buildings/party walls/rights of light surveyors
Archaeological survey	Local museum or British Museum
Environmental issues	Environmental consultant
Legal aspects	Lawyers
Outline planning permission	Architect
Safety issues	Health and safety specialist

The client's internal team should consider the main risks identified by the site investigations and in making the key choices identify options for dealing with each risk.

3.5 Standardized Solution or Original Design

Producing an efficient standardized solution or an innovative original design requires different work teams and ways of working. They provide clients with different facilities and levels of value and risk.

The first key choice is whether to buy a standardized solution or an original design. They provide quite different products and tend to be supported by different qualities of client service.

Once the choice is made, the project team must concentrate on delivering an efficient standardized solution or an innovative original design. The success of projects depends on being absolutely clear about which approach is being used. Different types of work teams are needed and they need to approach their work entirely differently using very different processes.

The differences include the way partnering is applied since good standardized solutions are most likely to be produced by project teams already experienced in using partnering. This may also be the case for original designs. However, the need to include diverse skills and knowledge tailored to the specific needs of the project and the need to involve the client in detailed decisions make it more likely that the project team will have to develop a distinct approach to partnering.

Many professionals in the construction industry believe that all projects are unique and they all require new answers. Designers in particular are good at identifying the potential advantages of original designs. For many projects they are right because no competent standardized solution exists. But this is not true for all projects and the construction industry has made considerable progress in developing ranges of standardized solutions that meet the needs of many clients. Five distinct levels of standardization are illustrated in Figure 3.2.

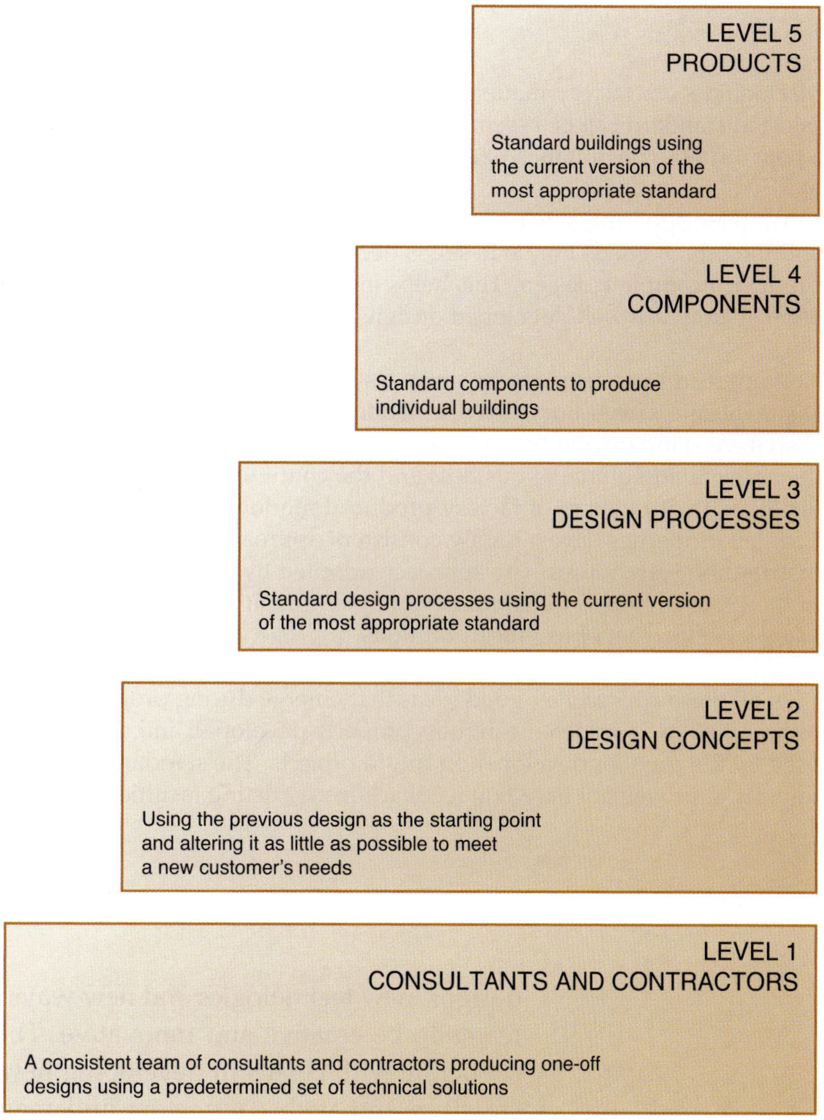

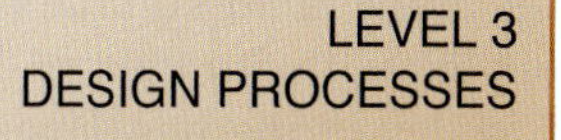

Figure 3.2 Levels of standardization

They should all be considered by the internal team before deciding to incur the greater costs, times and risks involved in original designs.

Standardized solutions are available for many types of buildings and infrastructure. They can provide reliable good value on simple, straightforward terms. In their most developed form, standardized solutions give clients low prices, fast delivery and reliable quality. In addition clients are likely to be offered sophisticated support services including help in finding sites, finance packages, show rooms that demonstrate the products, training for facilities managers, and maintenance and facilities management services. The questions in Chapter 7, checklist 2 will help clients decide whether a standardized solution meets their objectives. Only if it is certain that no standardized solution meets the client's objectives is it sensible to commission an original design.

Original designs are more expensive, take longer, are less predictable and involve clients in many more decisions. They require consultants, contractors and specialists with distinctive skills in producing new

Examples of Standardized Solutions

McDonald's use factory-made modules to produce their fast food outlets. The current version of the selected standard can be assembled on site in just a few days, thus allowing the income stream from a new outlet to begin much earlier than with traditional methods.

Waitrose supermarkets use a range of standardized design processes. Once a new site is available and the mix of goods they will sell is decided, computer-based design processes automatically produce a complete design. This helps individual projects to be undertaken quickly and efficiently because they use well-developed designs.

Whitbread have developed generic designs for the various types of buildings they need for their leisure business. Sainsbury's have a standardized approach to the main elements of their supermarkets. Esso have standardized the design of their service stations so that the only important variables are the anticipated volume of business and the configuration of the site. This approach has been further developed with the use of factory-produced modules for major elements of their service stations. Gazeley Properties have a highly consistent approach to the design and construction of large distribution warehouses. The approach adopted by Stanhope for major office buildings at Broadgate in the 1980s provided for continuous improvement from phase to phase within one mega-project using a defined set of technical solutions.

In all these approaches, good ideas that emerge during projects are generally not used on that project, but are considered carefully, properly developed and added to the standardized solutions used by the client or developer on future projects. The standardized solutions are also subject to continuous improvement based on feedback from existing facilities and research and development work.

designs, new technologies and new ways of working. They need work teams to be creative and innovative. They need work teams able to cooperate with clients in developing new kinds of buildings or infrastructure and with contractors and manufacturers in devising new forms of construction. The best original answers provide wonderful buildings and imaginative infrastructure. The best consultants, contractors and specialists specializing in original answers provide a realistic certainty about the project's outcomes by managing risks and working to carefully defined budgets and programmes.

3.6 Form of Project Organization

The construction industry uses different forms of project organization. The essential choice is whether the client employs one firm to take overall responsibility for their project or employs separate consultants, contractors and specialists to take responsibility for specific aspects of their project. The choice affects the time and resources clients have to devote to their projects.

The second key choice determines the main features of the project team's overall organization. The heart of this choice is whether to contract with a single organization or to employ a number of separate consultants, contractors and specialists to produce their building or infrastructure. Chapter 7, checklist 3 will help clients understand the main approaches used by the construction industry.

Using a single organization is simple. Provided the contract to employ them is well drafted, there will be no ambiguity over responsibilities. All the decisions and problems arising from any aspect of design, manufacturing and construction are the responsibility of the single organization.

However, before it is safe to give responsibility to a single organization, the client needs a clear and certain description of the building or infrastructure they need. They need to decide what they are willing to pay for it, and when they need it. Provided the client can make these decisions and then stick to them, they may well be best advised to employ a single organization to take responsibility for producing the building or infrastructure.

Design build

Construction firms who provide this single point of responsibility are usually called design build contractors. They undertake the complete project on the basis of information provided by the client's internal team. Producing sufficiently robust information and agreeing the terms of the contract with the design build contractor usually means the client's internal team needs the help of consultants.

The main approaches to design build are described in Chapter 7, checklist 3.

A decision to employ a design build contractor gives the project sponsor important responsibilities even though these are relatively small compared to alternative approaches. These responsibilities include dealing with the administration of the contract, helping to resolve problems and getting answers to questions raised by the client. The project sponsor may need professional advice for some aspects of this work.

It is sensible for design build contractors to use partnering in their project teams. However, if the selected design build contractor does not already use partnering, they are unlikely to be persuaded to change their approach on the basis of one project and beyond recommending them to read this code of practice, there is little point in the project sponsor taking this issue further.

Prime contracting

A single organization can be given responsibility for producing an original design. It can be an effective approach, especially if the client is sufficiently experienced to produce a detailed description of the performance to be achieved in the building or infrastructure and will not alter it during the project. This may well be expressed in terms of what the facility must do. This can be as simple as to provide accommodation for the organization's 500 design engineers in a building that encourages and supports the creative and efficient design of new products. Or provide a four-lane toll motorway over a defined route providing access points at defined intersections. Best practice takes account of the total life of the building or infrastructure by stating limits and constraints that apply to operational and maintenance costs rather than concentrating on capital cost alone.

Construction organizations that take single-point responsibility for producing an original design are called prime contractors. There are several different kinds of organizations able to lead an integrated construction supply chain including design build contractors, management contractors, multi-discipline design consultants, project management consultants and groups of firms combined in joint ventures. The prime contractor takes responsibility for integrating the design, manufacturing and construction activities as described in Chapter 7, checklist 3.

There is a strong case for using partnering throughout the prime contractor's supply chain. This should ensure that all members of the supply chain concentrate on achieving the client's objectives not least because they are rewarded fairly for their work on the project. There is also a strong case for using partnering in the relationship between the client's internal team and the prime contractor. Its emphasis on cooperative teamwork is exactly right for two organizations cooperating in the production of an original design.

Separate organizations

All modern construction projects involve many specialist professions, crafts and trades that are provided by separate organizations. This often makes it difficult for the client to be certain about the building or infrastructure they need, what it is sensible to pay for it, or when it is reasonable to expect it to be complete. Even with good advice the client may have significant doubts about some of these issues, so it is not possible to produce a robust basis for a contract with a single organization. As a result many clients, including in particular those that want an original design, usually employ a number of separate consultants, contractors and specialists.

Employing several separate organizations to design and construct the building or infrastructure gives clients the advantage of several independent opinions about key design, time and cost decisions. Project partnering reinforces this advantage by ensuring that all the various points of view are considered and decisions are well founded.

In broad terms the decision to appoint separate organizations gives clients a choice between using:

- A general contractor approach, which means employing consultant architects, engineers and quantity surveyors to direct the work of a general contractor.
- A management contractor approach, which means employing a design team and a management team headed by a management contractor.
- A construction management approach, which means employing a design team, a management team headed by a consultant construction manager and specialist contractors.

If the client decides to use separate organizations, the client's internal team will need good professional advice to help assemble the information needed to make decisions about the kinds of consultants, contractors and specialists that should be employed. It may be sensible to include a project manager amongst the advisors to guide the client in deciding which approach to use. Chapter 7, checklist 3 provides information about each main approach to help clients understand and evaluate recommendations from the internal team.

3.7 Tough Contracts or Partnering

Some clients prefer to use tough contracts and rely on everyone looking after their own interests. The results are often disappointing and many now use partnering. It requires more time and resources to work in cooperation with consultants, contractors and specialists but in return provides more benefits for everyone involved.

The third key choice is whether to use partnering. This choice arises because within current practice clients can use tough contract conditions to define responsibilities and liabilities and rely on everyone looking after their own interests. The alternative is to use partnering. This is likely to involve the client more fully in their project but in return provide a good chance that they will get a better building or infrastructure at a lower cost more quickly. Chapter 7, checklist 4 provides questions designed to help clients decide whether to use partnering or rely on tough contracts.

It is sometimes argued that the client can use tough contracts to provide a safety net to fall back on if partnering breaks down. Provided everyone involved agrees, there is no doubt that projects are undertaken on this basis and some are successful. However, using tough contracts and partnering puts the consultants, contractors and specialists involved in an ambiguous situation. They are in effect being asked to ignore the provisions in their contract and work as a cooperative team. The implication being that if partnering fails they can revert to the terms of the contract.

It is certainly the case that experienced professionals determined to make a success of project partnering make the form of contract irrelevant. However, there is a great risk that tough contracts based on the construction industry's traditional adversarial attitudes will force people into actions that substantially undermine partnering. Whatever the outcome, tough contracts and partnering do not mix well and project partnering is more likely to succeed if contracts designed for partnering are used. A number of standard forms of contract now make well-considered provisions for partnering and one of these should be used if a client wants to use partnering and also wants the security provided by formal contracts. Chapter 2 describes the standard forms of contract available.

3.8 Project Feasibility

The three key choices establish the feasibility of producing a building or infrastructure that meets the client's objectives.

The client's internal team should routinely check at key points throughout the project that it is feasible for a building or infrastructure to be produced that meets the client's objectives. These are inevitably spot checks, not a guarantee of success. Particularly in the early stages, many decisions have to be made and many actions carried out successfully before the project's actual achievements are known.

The first key point for a feasibility check is when the key choices have been made. In carrying out this feasibility check the client's internal team should review the overall strategy defined by the key choices they have made. They should consider any other equally good strategies. They should consider whether they have given enough consideration to opportunities for the client to use the new facility to change their business or the way they work. They should consider whether there are radically new forms of facility that could provide better answers. Then all strategies that appear to have the potential to provide what the client needs should be evaluated to determine the best approach and to make sure it is feasible.

The client's internal team will need help from the various advisors they have appointed to produce detailed evidence that the client's objectives are feasible. This should include evidence of the costs and times achieved on similar projects that were completed successfully. It should include evidence of the benefits delivered by such projects. It should explain how this previous performance relates to the client's project. It should include a review of the risks involved in the client's project and suggest how they can be managed. As the *Code of Practice for Project Management for Construction and Development* puts it, the feasibility stage ensures that a suitable site is available, and produces a project brief, a design brief, a scheme design and a funding and investment appraisal.

A common way of helping to ensure the feasibility of a construction project is to include contingency allowances in the budget. In a sense this legitimizes failure. Some project teams recognize this by replacing the contingency allowances with lists of savings that can be made if problems make them necessary and of extra things they can include if the project goes well. Balancing savings and opportunities is a move in the right direction but this still gives project teams an easy way of resolving problems at the client's expense. Many experienced clients refuse to allow project teams to include any contingency allowance in their cost plans. Instead they give their internal teams contingency allowances that are kept secret from the project team. In practice it becomes obvious that a contingency allowance exists as problems arise and are dealt with by the internal team finding extra money rather too readily. These various ways of providing contingency allowances are often used when a project team is in the early stages of partnering. They do not provide the best approach.

Best practice is to have no contingency allowance. It is better to use value and risk management as described in Chapter 7, checklists 29 and 30 in producing robust plans for the project. The point of this tough approach is to ensure that the client gets the maximum possible value for all the money justified by the business case for the project. It challenges the project team to aim at the exceptional performance of delivering the building or infrastructure exactly on budget.

Many people in the construction industry regard the idea of having no contingency allowance as hopelessly unrealistic. However, there is growing evidence that construction projects that aim high achieve excellent performance. Aiming high means consistent, hard work but everybody gains when project teams fully accept the highest feasible objectives.

3.9 Selecting the Project Team

The client's internal team selects consultants, contractors and specialists to provide work teams that form the project team. The first firms appointed are those that provide members of the core team. This small team of key individuals gives the project its overall direction. They guide the appointment of all the other work teams needed.

Once the key choices have been made and the project's feasibility checked the client's internal team begins selecting consultants, contractors and specialists to provide the work teams that form the project team. The guidance in Chapter 2 should be used to ensure that work teams have the essential technical skills and knowledge and a partnering ethos. Chapter 7, checklist 33 provides a glossary that describes the various kinds of teams used in modern construction.

The first work teams to be appointed should include key individuals who with the project sponsor will form a core team. The core team provides the project's overall direction. Members of the core team should be experienced professionals in the relevant business, design, manufacturing, construction, regulatory and political issues.

Core teams should be small – three or four people are ideal – and a practical maximum is seven. It is therefore likely that from time to time they will need expert advice from outside the core team. This may come from a firm appointed to provide part of the project team or from external experts commissioned to provide specific advice or undertake a short-term study.

The core team guides the appointment of all the work teams needed to undertake the project. In addition to ensuring that decisions about the design, manufacturing and construction activities are consistent with the client's objectives, the core team takes the lead in ensuring that the agreed approach to project partnering is put into effect.

An important practical issue is that construction projects require the work of hundreds of specialist work teams. Some of these contribute to the project for only a short time. For these reasons it is impractical to involve all the work teams fully in project partnering. The answer is to recognize the existence of technology clusters. That is broad groups of related technologies that provide the major elements or systems of the building or infrastructure. In building projects typical major elements and systems are:

- Substructure including underground services
- Structure
- External envelope
- Service cores, risers and main plant
- Finishes and services to entrance and vertical circulation spaces
- Finishes and services to horizontal spaces
- External works.

Infrastructure projects similarly give rise to distinct technology clusters comprising firms that specialize in major elements or systems of the end product. Within each cluster the technology dictates a natural structure of roles and responsibilities that form a supply chain. This natural structure includes lead firms able to represent the interests of their supply chain in project teams.

One of the criteria for selecting lead firms should be that they use partnering in working with the other firms in their supply chain. Lead firms should be appointed early and fully involved in project partnering. Indeed, when their element or system is central to current project decisions, the lead firm should provide a member of the core team. The practical arrangements needed for this to work effectively are described in Chapter 4.

Project partnering requires members of the project team to be appointed as early as possible. When it is decided that the best approach is to use a standardized solution, the core team can appoint the whole project team at the start of the project. Original designs often give rise to unusual and unexpected technical issues that cannot be anticipated early in a project. Therefore the core team identifies the technical knowledge and experience needed as early as possible and then appoints suitable consultants, contractors and specialists. Once the initial appointments are made the first partnering workshop should be held.

3.10 The First Partnering Workshop

Project partnering is guided by partnering workshops. The first partnering workshop shapes the project's inputs, processes and outputs by agreeing mutual objectives, decision-making processes and performance improvements.

The way project partnering will be applied is reviewed, clarified and agreed at the first partnering workshop. This is normally a two-day meeting of the project team at a neutral venue run by a partnering facilitator.

Inputs	Processes	Outputs
Cost	Risks	Capital value
Time		Operating efficiency
Quality		Life-cycle costs
Safety		Environmental impact
		Profits
		Experience

Workshops have become a crucial part of a number of widely used management techniques including value management, risk management and team-building. These techniques supplement established project management methods and show that something more than management is needed for construction projects to deal with the complex issues that clients, technology and the wider community throw up. Partnering goes much further in using workshops to radically alter the way projects are undertaken. Partnering uses workshops to shape all the project inputs, processes and outputs by concentrating on three sets of primary decisions:

- **Mutual objectives,** which define the project ***outputs***
- **Decision-making,** which shapes the project team's ***processes***
- **Performance improvement,** which aims to reduce the ***inputs***.

The box above provides a quick checklist of issues that project teams should consider in identifying and agreeing a set of actions that will ensure their project's success.

In addition the first partnering workshop makes sure the project has good feedback systems. They turn the basic model of inputs, processes, outputs into a controlled system. Feedback is an important aspect of the construction industry's work that is often neglected, which is why it needs to be explicitly considered at the first partnering workshop. More detailed guidance on the organization of partnering workshops is given in Chapter 7, checklist 19.

The first partnering workshop takes two days to allow participants to consider a great deal of information, generate ideas and make tentative decisions on the first day. Overnight all this is worked on subconsciously as they sleep. Next morning at least some of the participants will see new ideas, better answers and everyone is likely to feel more confident in making decisions. Two-day workshops build on the basic human phenomenon that our brains continue to work while we are asleep. It may be possible for a workshop held on one day to be effective if the team has worked together before on similar projects but in general, the first partnering workshop should take two days.

3.11 Mutual Objectives

The first partnering workshop agrees what the client, consultants, contractors and specialists will get from the project. This is achieved by searching for win–win agreements using value management techniques. The resulting mutual objectives are further developed following the workshop into precise descriptions of the project outputs.

The first main task for the first partnering workshop is to agree the project's mutual objectives that specify the value delivered to the client and the profits earned by consultants, contractors and specialists. This does not mean that the only important outputs are financial. The project's main output will be a building or infrastructure that may well enable the client's organization to operate more efficiently, provide a better service to its customers, provide healthier working conditions for staff and delight neighbours. The workshop takes account of all these capital, life-cycle and operating issues in defining the value delivered to the client.

Similarly the project may provide important lessons, experience and contacts for the consultants, contractors and specialists involved. These non-financial benefits should be taken into account as well as the profits generated by the project. In other words all the potential benefits should be considered before defining the value and profits the project team aims to deliver.

It is nevertheless important to agree financial arrangements that encourage everyone involved to concentrate on the overall success of the project. It is energizing when people agree that they want to make money and understand the only way to do this is to cooperate with each other.

Early use of partnering often relied on profit-sharing schemes but it proved difficult to avoid these degenerating into disputes when problems arose. It is much simpler to guarantee the cost to the customer of a defined product and to guarantee each construction firm an agreed, fair profit plus all their properly incurred costs. This financial security allows everyone to concentrate on doing their best work and completing the project to the very best of their ability. The general nature of these arrangements is described in Chapter 7, checklist 26.

These mature financial arrangements support a growing understanding in the construction industry that long-term success depends on win–win agreements. Construction projects can provide sufficient profits for everyone to have all they need if they work together to create it. Win–win is based on the idea that 'There are better ways of working than my ways or your ways if we take the trouble to look for them. The better ways enable everyone to win.'

It is therefore sensible for first partnering workshops to keep talking until agreement is reached on mutual objectives that everyone can fully accept. Indeed if this turns out to be impossible, it may well be that partnering is the wrong approach for the project or the particular project team. Mutual objectives that give everyone more than they could get from concentrating narrowly on their own interests are fundamental to project partnering.

In searching for mutual objectives, it is unhelpful for people to have fixed, predetermined answers. This can be avoided by getting everyone at the workshop to describe their own interests in specific, practical terms. It helps if they talk in detail about what they want from the project and how it will help their business. An essential basis for good mutual objectives is that all parties are clear and open about their own interests. Getting people to describe what they want in front of the rest of the team helps ensure they are reasonable in what they ask for. It often helps for people to state what they see as an ideal outcome, what they reasonably hope for if the project goes well and their minimum requirements.

Once everyone at the workshop has described their own interests, the next step is to generate a range of possible mutual objectives. The aim is to look for solutions that satisfy as much as possible of everyone's interests. This is helped by using creative techniques like brainstorming, adopting the point of view of different professions, or suggesting newspaper headlines to describe the outputs. Working together creatively can produce a relaxed frame of mind that opens the discussion to a search for answers that provide mutual benefits that meet everyone's vital interests and can give them more than they expected. This approach draws on techniques developed in value management based on the general framework in the box below which is explained in detail in Chapter 7, checklist 29. Also the Bibliography includes publications on value management.

The agreed mutual objectives should be further developed in the weeks following the first partnering workshop into a detailed statement. This should define the agreed outputs in terms of what is to be done and by when; the constraints within which the results are to be achieved; the standards to be used in evaluating the outputs; the timing of evaluations; and the actions that will flow from evaluations.

Value Management Framework

- What are we trying to achieve?
- What must we get right to achieve it?
- What constraints apply that we must take into account?
- What is the relative weight of each of these factors?
- How do the available options contribute to achieving our aims?
- How can more value be delivered?
- Which are the best ideas for adding value?
- How can they be implemented?

3.12 Decision-Making

The first partnering workshop agrees the processes and organization framework to be used in running the project. This establishes how decisions will be made and problems resolved. It often includes risk management techniques. It provides procedures and standards, identifies constraints, establishes targets and control systems, and ensures that information systems support the agreed approach.

The second main task for the first partnering workshop is to agree the processes to be used in running the project. The agreement should provide an organizational framework for making decisions and guiding the behaviour of the people involved. The overall structure of decision-making systems is described in Chapter 7, checklist 8 and is illustrated in Figure 3.3. All the elements shown in Figure 3.3 should be considered at the first partnering workshop.

In deciding on the specific decision-making system to be used, the broad aims should be to concentrate the project team's efforts on achieving the agreed mutual objectives, to ensure that few problems arise and those that do are dealt with in ways that do not threaten the partnering relationship.

The first partnering workshop needs to agree how decisions will be made. It is particularly important to agree how the client's internal team will be involved so they are able to influence all decisions that may affect the value delivered by the end product. This is particularly important when an original design is needed. It is also very important

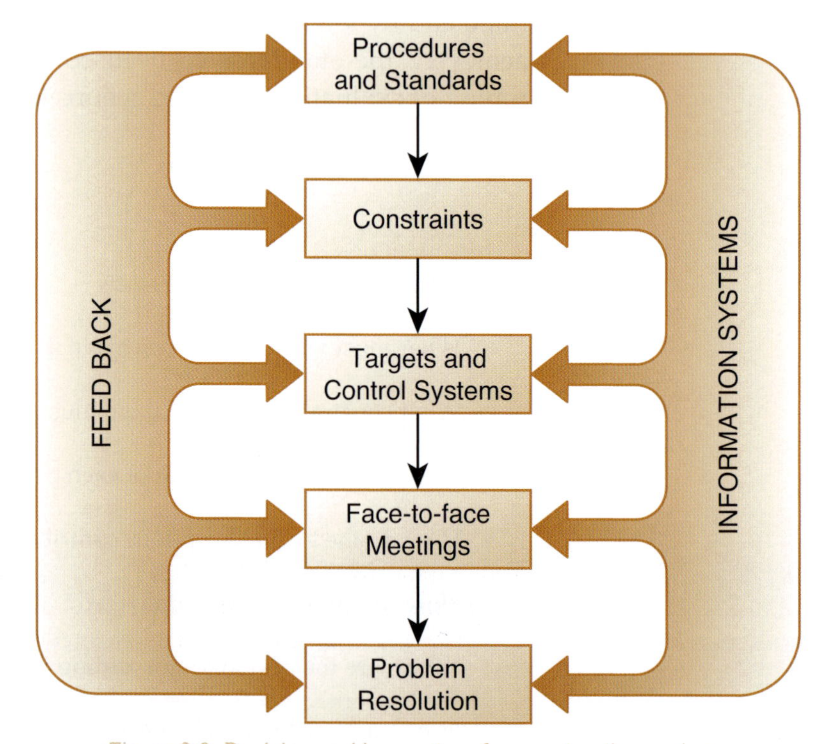

Figure 3.3 Decision-making system for construction projects

to agree procedures that ensure the client's internal team is fully informed about all aspects of progress. The aim should be that they do not have any negative surprises.

It is also important to agree where the project team will be based. Traditionally people work in their own firm's offices. This is cheap but tends to inhibit communications between work teams and cause people to fall back on routine decisions rather than search creatively for a better answer. Recognizing these limitations, a growing number of successful projects set up a common project office where all the professions and specialisms can work closely together. This makes communications fast and accurate, allows people to discuss innovative ideas and encourages open decision-making. There are costs involved in setting up and managing a common project office and in people working away from their home base. These should be balanced against the potentially large benefits from a physically, intellectually and emotionally united project team. Chapter 7, checklist 3 provides more information about the use of a common project office.

Having considered where the project team will be based, the workshop should agree what decision-making tools they will use. The first type of decision-making tool is procedures and standards. Procedures are predetermined actions for workers to take in given situations and so in effect procedures define standardized processes. They are closely linked to standardized products and services, commonly referred to as standards. Many well-established procedures and standards exist and save time and resources by providing predetermined answers to many of the issues that face project teams. It is normal for work teams to have their own preferred procedures and standards and so the first partnering workshop should ensure these are mutually compatible.

In doing this it has to be remembered that procedures and standards may be formally written and approved by firms, industry bodies or government. Or they may merely be implicit in teams' work. In whatever form they are recorded, procedures and standards tell teams how to undertake their work and how others should behave in given situations. Knowing other teams' procedures and standards gives everyone the confidence to concentrate on doing their own best work because they understand what other work teams are doing and why. As a result they know what needs to be done for the project as a whole to be effective. The use of procedures and standards in partnering is described in Chapter 7, checklist 9.

Procedures and standards usually deal with a range of situations and so teams need to determine the precise constraints that apply to their particular work. It is worth spending a few minutes at the first partnering workshop ensuring that there is a common understanding of the constraints that define the levels of performance that have to be met. Construction constraints include official regulations about the forms of construction that are permitted, the methods that can be used, safety requirements and quality standards. Best practice makes constraints an integral part of processes but where the work is relatively new to the teams undertaking it, they may need control systems to ensure they are working within all the constraints that apply. The use of constraints in partnering is described in Chapter 7, checklist 10.

The targets and control systems that teams use in organizing their own work are another essential part of project decision-making. Targets provide a measure of performance that teams aim to match or beat. The existence of targets means teams require control systems. At their best, control systems rely on teams aiming at their target as an integral part of their work. The nature and extent of the control systems needed depends on the team's normal performance and how close the target is to this norm. When a target is comfortably within a team's normal performance, control can be simple and infrequent. When a target is challenging, teams need detailed control systems that work in real time to provide feedback that tells them how close they are to meeting their targets.

The first partnering workshop should review the targets and control systems used by each of the work teams to ensure they are compatible with each other and with the agreed mutual objectives and performance improvements. It is normal for there to be gaps and overlaps between the normal approach of the work teams that make up a project team so the workshop needs to agree overall targets and project control systems. The use of targets and control systems is described in Chapter 7, checklist 11.

It is particularly important for the first partnering workshop to agree how the project team should deal with crises or problems. The agreed approach may include informal meetings, formal meetings, workshops or task forces. The nature, timing and form of these should be agreed. Chapter 7, checklist 12 describes effective patterns of meetings for projects using standardized solutions and checklist 13 describes those for projects producing original designs. The use of task forces in partnering is described in checklist 22. The use of workshops in partnering is described later in this chapter.

These arrangements for dealing with problems and crises should be based on the fundamental principle that people must look at their own responsibility and not blame others. Solutions come from everyone involved concentrating on what they can do to help solve the problem, not worrying about what others should do. Focusing on what other people should do creates conflicts, implies blame and produces suboptimal solutions. Arrangements that support effective crisis- and problem-resolution techniques are described in Chapter 7, checklist 16. The aim in dealing with problems and crises should be to find permanent answers that allow the project to resume normal planned and controlled work quickly.

In considering these general arrangements for dealing with problems and crises, the workshop should ensure that problems caused by individuals who do not accept the disciplines of cooperative teamwork are dealt with. It is important that the whole team recognizes that such behaviour must not be accepted. This means that it is accepted as normal for anyone faced with an uncooperative team member to discuss it with the offending individual face to face. If this

Risk Management Framework

- What is at risk and why?
- What are the specific risks and where and when could they occur?
- What are the consequences of each risk if it occurs?
- What is the likelihood of each risk occurring?
- How will each significant risk be dealt with?
- What specific actions need to be taken?

fails, it must be raised as a problem and dealt with using the agreed procedures. A range of solutions including coaching, mentoring, training or replacing the individual should be provided.

The various tools described above are supported by information systems which provide the final decision-making tool for the workshop to consider. Good information systems ensure that everyone involved with the project in whatever capacity is using current information in a convenient and relevant form. Information systems feed into and draw from the other decision-making tools to provide the feedback that gives project teams an essential basis for controlling their projects. The use of information systems in partnering is described in Chapter 7, checklist 14 and feedback in checklist 15.

The workshop should ensure that the overall framework of decision-making is consistent with techniques developed in risk management and takes account of the general risk management framework in the box above. More information about risk management is provided in Chapter 7, checklist 30 and publications listed in the Bibliography.

3.13 Performance Improvements

The first partnering workshop agrees one or two specific performance improvements. Agreed improvements often aim to increase the value delivered to the client. The workshop agrees how the improvements will be measured. It agrees actions to achieve the improvements which are further developed following the workshop.

The third main task for the first partnering workshop is to identify the one or at the most two performance improvements they will make. This requires them to make specific savings in the inputs used in achieving the agreed outputs using the agreed processes.

Savings in inputs should be related to some measurable standard. This may be the normal performance achieved by the local construction industry, the team's previous best performance or some other agreed standard.

In undertaking this part of its work, the first partnering workshop should keep in mind that the whole point of partnering is to deliver performance improvements. In recognition of this the first partnering workshop should spend a significant proportion of its time agreeing the specific performance improvements they will achieve. These must be consistent with the agreed mutual objectives and be clearly reflected in the targets the project team set for themselves. Chapter 7, checklist 11 suggests areas of project teams' performance that could be the subject of improvement targets.

A good starting point for agreeing a specific performance improvement is to identify a problem that limits the value delivered to the client. This is because the long-term health of consultants, contractors and specialists is benefited more by solving problems that inconvenience clients than by tackling internal problems. In practice delivering better and more reliable value for clients usually requires some internal processes to be improved. The point is that the long-term interests of all construction firms are best served by internal improvements likely to have the largest beneficial impacts on the value delivered to clients.

Once a problem is identified, the sequence of actions which gives rise to the problem and its consequences should be analysed. This means identifying for each action all the inputs, constraints and outputs. The purpose of the analysis is to help identify the causes and effects that determine performance. The general pattern of this analysis is illustrated in Figure 3.4. Having identified the key features of the actions surrounding the problem, the next step is to measure them so as to establish the extent and location of the problem. The measurements also help set a measurable target for improving the existing situation. This should be an ambitious target set by the project team itself.

Following the workshop a nominated individual or task force should define specific actions designed to achieve the planned improvement. This may need help from experts or researchers from inside or outside the firms involved in the project. Possible answers should be discussed with everyone in the project team likely to be affected by possible changes. The aim is to agree a robust way of achieving the agreed performance improvement that the team feels confident about.

Benchmarking provides a well-developed approach to finding ways of improving performance that can be used in partnering workshops and the follow-up actions. Chapter 7, checklist 31 provides guidance on benchmarking.

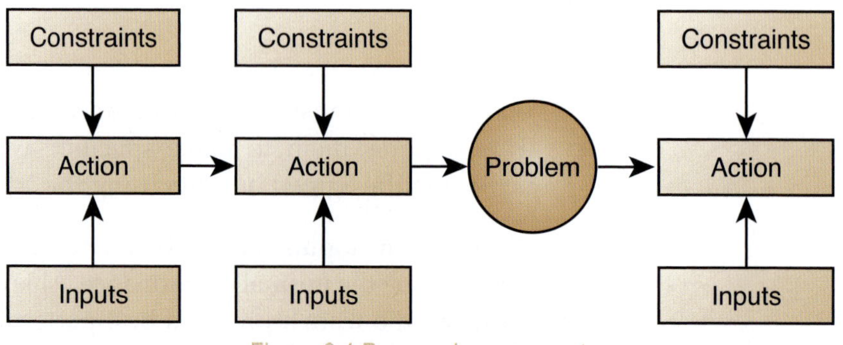

Figure 3.4 Process Improvement

3.14 Feedback

The first partnering workshop agrees feedback systems to provide work teams with reliable information on progress and early warnings of problems. It also establishes the basis for lessons to be captured for use on future projects.

All controlled systems, including construction projects, depend on feedback. Feedback is crucially important in enabling teams to achieve their mutual objectives and deliver agreed performance improvements. It enables teams to exercise control on the basis of their actual outputs rather than on their planned performance.

It is therefore important for the first partnering workshop to check that the agreed processes provide robust feedback. The most effective feedback comes when teams measure their own performance and compare it to their targets. Depending on the outcome the project team can:

- Continue with the same actions because they are producing acceptable results.
- Make changes to bring performance back into the range of acceptable answers.
- Set a new and more ambitious target if the project is going exceptionally well.

The workshop needs to recognize that there are considerable differences in the timescale of construction feedback. There is a great deal of immediate feedback. The craftsman working with a material obtains second-by-second feedback on the effect of his actions. He can see the effect of his actions on the material and he can feel the effect through his hands as he works. However, feedback on the performance in use of the part he is making will take time to become available. Similarly, an architect sketching the first designs for a building may have to wait years before his actions are joined in a feedback loop by knowledge about the appearance of the actual building.

Project teams are generally concerned with short-term feedback; that is, feedback which arrives in time for it to influence their behaviour while they are still working on the project. Longer-term feedback is valuable in that it provides knowledge which helps future projects. This is an important reason for construction to use well-established answers wherever possible. Standards and procedures developed on the basis of feedback over many projects provide a robust basis for new projects.

When teams depart from well-established answers, it is especially important to ensure they have reliable feedback to provide early warnings of problems. This is crucial in helping them know where to concentrate their efforts. Chapter 7, checklist 15 provides advice on effective feedback systems.

Project teams should regularly report their achievements to senior managers so they know partnering is delivering real performance improvements. It is vital that this includes senior managers in the client's organization so that chief executives in all sectors of the economy can speak confidently about the success of their construction projects.

With good decisions about feedback in place, the workshop can reasonably expect the project to behave as a controlled system that achieves their mutual objectives. Also they will be well-prepared for the final partnering workshop, which captures lessons for use on future projects.

3.15 Workshop Outputs

The decisions made at the first partnering workshop can be promoted in a partnering charter produced at the workshop. The decisions also provide the basis for the Project Execution Plan and Project Handbook. These key control documents are produced following the workshop and kept up to date throughout the project.

The first partnering workshop should review its main decisions to ensure they are mutually compatible. A sensible check is to consider whether the workshop's decisions will result in too many meetings. This may indicate that too many new decisions are required and the workshop should reconsider the potential benefits of new decisions against the established benefits of using good standardized solutions. The box below lists key characteristics of the decisions for projects using a standardized solution or an original design.

	Standard Answer	Original Design
Mutual Objectives	Well-established design delivering predictable value for the client and certain profits for the consultants, contractors and specialists.	An innovative design that will take account of everyone's interests and use value management techniques to deliver exceptional value for the client and guarantee consultants, contractors and specialists' profits.
Decision-Making	Predetermined structure of relationships, meetings and processes that take account of the risks likely to be faced.	A structure that develops in response to project decisions guided by a strong core team and an overall pattern of meetings. The project team uses a wide range of decision-making tools flexibly, including risk management.
Performance Improvement	Use of inputs that match current best performance plus one specific improvement developed by a task force outside of any project.	Use of inputs is based on current industry norms plus one specific improvement agreed by the project team.
Feedback	Well-established and effective feedback within projects and from project to project.	Feedback systems are tailored to the needs and methods of the work teams that form the project team. They are reviewed regularly and if necessary adjusted, especially when new teams are appointed.

The *Code of Practice for Project Management for Construction and Development* describes two key control documents. First is the Project Execution Plan, which states the authority and responsibilities the client has given the project team. It is described in Chapter 7, checklist 27. Second is the Project Handbook, which describes how the project team will work together. It is described in Chapter 7, checklist 28. The early stages of projects culminating in the first partnering workshop provide the information needed to produce the first versions of these two key control documents. They should be the minimum essential statements to guide the project team. They should not be treated as formal legal documents with every word and line haggled over for individual advantage. They are notes to ensure everyone knows what they have agreed and are developed throughout the project as decisions are made.

Ideally all decisions will be agreed at the workshop but on many projects this is not possible. In these circumstances the workshop should agree how and when final decisions will be made. It may need to define further work for specific individuals, arrange special meetings to consider unresolved issues or set up task forces to tackle difficult issues. Sometimes the best course of action is to reconvene the workshop within a few days or at the most a few weeks. The workshop should agree a timetable for the follow-up actions stating who is to do what by specific dates. The need for follow-up actions should not be interpreted as a failure by the workshop; it simply makes sense to take the time needed to develop proposals thoroughly.

In addition to the decisions directly relating to the project, the workshop may identify a need for training. With firms new to partnering this may include training in cooperative decision-making and teamwork. On projects facing tough targets, this may include training in the application of quality control systems or cost control systems based on open book accounting or flexible time control systems with fixed completion dates. Particular safety concerns may give rise to the need for training. On projects that have aroused the interest of the media for whatever reason, it may be necessary to provide training in public relations.

When the decisions have been made they should be recorded. Good practice includes preparing a clear, punchy statement of the agreed basis for project partnering and publishing it as what is often called the partnering charter. This can be displayed in offices and site accommodation, presented on cards that can be kept in a wallet, and used as an introduction in electronic information systems to remind project team members of what they are cooperating to achieve and how they have agreed to work.

Finally, it is sensible to consider whether the project may benefit from an independent review of its approach and performance. It can be very beneficial to commission an ongoing case study during the early stages of a project. Then interim results can inform milestone workshops and the final results provide an important input to a final workshop. There is a growing number of well-researched case studies of partnering projects that help define and establish best practice. They also help project teams develop by identifying weaknesses and problems.

3.16 Organization of the Workshop

The first partnering workshop is organized and guided by a partnering facilitator who ensures that all key decisions are made. Everyone able to influence the project's performance should attend.

The guidance about who should attend and the essential preparation in Chapter 7, checklist 19 applies to the first partnering workshop. It is particularly important that the client is represented by senior managers who understand exactly what the project has to deliver.

The workshop has important decisions to make but it makes sense to use the first morning to get people talking about themselves, their interests and the way they like to work. This may be helped by using character and personality tests and team-building games.

The initial introductions are important but experienced facilitators know that teams develop by working together and so they move quickly into the real issues. This is achieved by the facilitator leading the initial session into a discussion of what each organization wants from the project and what they hope to gain from using partnering. This crucial part of the workshop should be well under way by the first afternoon. It should produce well-considered statements of what each participant wants from the project for them to consider it a success. These statements should be used to identify any problems or conflicts between individual expectations and needs. Then the workshop should agree how these can be resolved.

Then, still in the first afternoon, the workshop should discuss, argue about, explore and agree mutual objectives and specific performance improvements. These should be defined as precisely as possible so it is clear how performance towards each objective and improvement will be measured and specific targets set.

These important agreements are often debated long into the evening. This allows everyone to understand the implications of what they have agreed, spot problems and develop better or different ideas. Having slept on all this, the first session on the second morning is crucial. The previous day's decisions should be reviewed and everyone present given the opportunity to say that they still agree, have identified problems they believe can be resolved, or have identified problems that means the workshop should think again.

The aim should be to have mutual objectives and performance improvements agreed by the middle of the morning. There may be some remaining reservations that need to be dealt with by discussions and meetings in the days or weeks following the workshop.

The workshop next concentrates on agreeing the decision-making processes the project team will use. Much of the detail should have been discussed and agreed before the workshop but the workshop now provides the opportunity to review how they will work together. It often helps for the workshop to break into small groups at this stage to discuss specific aspects of the decision-making tools. Progress can be made quickly in this way and discussions at the plenary reporting back session help build commitment to agreed ways of working.

The need for task forces to deal with any aspects of the project should be considered after lunch on the second day. These may include a task force to plan how the agreed performance improvements will be achieved. It may include a task force to look at specific technical issues concerning the client's use of the building or infrastructure, design problems or possible construction methods. It is sometimes possible for a task force to have a preliminary discussion during the workshop but it is more usual for them to start work a few days afterwards.

The final action is to agree the partnering charter. The facilitator will produce and regularly update a draft as the workshop makes decisions. It helps if the current version is displayed in the main workshop room so everyone has a chance to discuss it during breaks throughout the workshop. If they can see improvements or think it misrepresents what they agreed, they should tell the facilitator. This can result in several alternative versions being displayed and discussed. This ongoing, interactive process should help the workshop agree the final set of words to form the partnering charter.

3.17 Cooperative Teamwork

The first partnering workshop provides an important opportunity for cooperative teamwork to develop and the facilitator should explicitly help and encourage this.

Cooperative teamwork is central to partnering so in addition to producing the formal outputs described above, the first partnering workshop should encourage cooperative teamwork. Experienced facilitators do this by using techniques and games that encourage cooperative teamwork which usually include some of those described in Chapter 7, checklist 19. These are used to generate excitement, break out of stalled discussions, focus participants' attention on decisions they need to make, keep everyone committed and encourage cooperative teamwork.

Many people's experience in construction makes it difficult for them to accept that cooperative teamwork is more efficient and will give them more benefits than concentrating on looking after their own individual interests. Partnering workshops help but for those that need more convincing, Chapter 1 describes important ideas that should help people recognize the benefits of cooperative teamwork.

Partnering for multiple projects

Case Study Reference: 020

Substantial cost and time savings on multiple projects worth £30 million were achieved by international civil engineering contractor Costain and its major client Thames Water, through a successful partnering contract, formed in 1997. The contract was for three years, with provision to extend to five years.

At the start of their partnering project, the first task for senior managers from Costain and Thames Water was to commit to the principle of partnering as a springboard for change. Once that commitment was in place it took about three months to integrate contractor and client staff. In planning the integration, the two partners made three major resolutions:

- To make the team's diversity of work experience and cultural backgrounds a benefit rather than an issue.
- To expand people's roles to give them a wider perspective of project management.
- To surmount the barriers of traditional contractor/client communication.

The next task was to ensure that team relationships at all levels were developed rapidly. This was achieved through a series of peer group workshops focusing on the adoption of best practice partnering and a training programme to develop team-building skills. Following this, a number of work groups were created to manage key stages of multiple projects.

Physical proximity of team members was considered crucial to the team's development and ease of communication. From the outset senior managers assigned to the partnering initiative by the client and contractor, along with commercial and accounting staff, were housed in a common office as one homogenous team.

The other main task in the early stages was to adapt management techniques and systems to the partnering contract. The changes included introducing new financial systems to enable open book tendering and cost monitoring. It took about nine months to establish the necessary changes.

The early benefits of the partnering initiative were:

- Accurate workload forecasts were made possible by developing multiple project programmes over entire financial years.
- Better project outcomes, because involvement of staff from both Costain and Thames Water at concept stage increased buildability and provided robust design solutions.
- Mutual trust and cooperation fostered by open book financial arrangements.
- Staff skills and work experience enhanced by being involved throughout the whole project cycle and then applying the lessons to repeat projects.
- Project estimates matched the outturn costs to within ± 5%.
- Contractor's workload and construction programmes were more secure.

Measured partnering

Case Study Reference: M4i 20

Three years after entering a partnering agreement with Hampshire County Council, highways contractor Raynesway Construction Southern won a further two-year contract on similar terms. Asked why his company was so successful in its relationship with Hampshire County Council, managing director John Jackson replied: 'We're an innovation-driven business offering clients the best value for money. Informed clients are not going to sign up to partnering deals just for the feel-good factor, they want evidence that performance will improve. We call it measured partnering.'

Raynesway Construction Southern and Hampshire County Council agreed eight headline key performance indicators that are aggregates of many process indicators. Raynesway Construction Southern's managing director John Jackson explained: 'It's not all one way in favour of the client. The winter maintenance response key performance indicator favours the client, while the invoice turnover key performance indicator is one we're particularly keen on.' Hampshire County Surveyor, John Ekins, reinforced the view that the key performance indicators are balanced: 'Actually, we're also interested in invoice turnover because predictability of cash flow is critical in managing our budget.'

Once measured partnering is secured as good practice throughout the industry, it will give clients the confidence to change procurement policies in favour of consistent good performers. Assistant county surveyor

Alan Mills argued: 'If contractors are winning typically one tender in four, then we're paying four times the cost of tendering for each contract. The industry needs to find a way to avoid this waste and measured partnering provides the best hope of doing just that.'

The benefits of measured partnering included:

- Cost predictability – the number of invoices submitted within 14 days of completing the work increased from 75% to 88%. A highways client needs to know the financial implications of work orders as soon as possible and this key performance indicator pushed the contractor into improving their paperwork.
- Invoice payment within 28 days jumped from 80% to 95% under the measured partnering arrangements. 'I'm delighted to see the capital employed falling,' Raynesway Construction Southern's managing director John Jackson confirmed.
- Fewer accidents – under the partnering arrangements, reportable accidents have reduced by 70%.
- Fewer defects – the number of defects was incredibly low, only 1.5% of the 8,000 work orders in a year required remedial work.
- Increased productivity – Raynesway Construction Southern and Hampshire County Council delivered about 17% more work with the same workforce.

30% programme gain by third Asda project

Case Study Reference: M4i Rethinking Construction Case History

Asda recognized a golden opportunity to achieve continuous performance improvements when they required a series of three similar supermarket projects in Scotland.

Asda's project managers, Capita Symonds, established a partnering arrangement with the key players in the project team. All the partners had already worked on Asda projects but the project managers decided to use partnering to cement them together as an integrated design and construction team. This led to partnering arrangements with the design team led by architect Percy Johnson and the constructors led by HBG Construction.

The partnering team's agreed aims were to deliver the programme and quality Asda wanted in exchange for continuity of work. They focused on three key principles from the Egan report's recommendations:

- zero defects at completion
- faster construction programmes
- harnessing the supply chain to make continuous improvements.

Actions at the start of the project included implementing a shopping basket of value-engineered solutions developed previously in partnership with Asda. Led by project managers Capita Symonds, the team identified innovations that provided radical improvements in the construction programme. The final store in the programme was constructed in 30% less time than the first one. For Asda, that meant eight weeks' extra sales income and for the construction supply chain, an accelerated cash flow.

Also the team concentrated on quality control. A key part of this was making all the design and construction teams, as well as Asda's facilities management and user teams, jointly responsible for verifying that the project was free from design and construction defects. As a result the supermarkets were virtually defect-free at handover.

Thames/Morrison network partnering pioneers

Case Study Reference: M4i 73

Thames Water Utilities and Morrison maintain the vast South London water supply network using an innovative partnering arrangement. They share workload forecasting, operational resources and facilities, as well as keeping open book accounts. Payment to Morrison is based on performance against agreed target rates established at the start of the partnering arrangement.

Nick Hester, Thames Water Utilities' Customer Services Director, is delighted with the results: 'Customer satisfaction and regulatory targets are key focus areas which this style of partnering helps us deliver'. His colleague Andy Hall, who manages network service providers, agrees: 'This achievement enables us to work with Morrison to squeeze out waste so we can cut our unit costs without compromising service levels.'

Morrison Area Director Adam Gosnold is enthusiastic about the effort the partnering team put into monitoring performance: 'It is clear that successful partnering requires joint performance objectives to be measured. We measure the performance of the work in a number of ways and this in itself has led to significant improvements in performance.'

Significant results were produced in the first two years. Repair time fell considerably, there was a dramatic reduction in the number of defects and overall predictability was much improved. The key measures included:

- 30% improvement in job completion on time
- 32% better permanent reinstatement compliance
- 65% improvement in meeting job priority timescales.

Actions during projects

Planning for Partnering

Involving the Teams

Reinforcing Partnering

4.1 Introduction

This chapter describes actions taken during the main stages of construction projects by clients and construction teams using project partnering. The actions are organized in distinct stages marked by milestones.

Partnering works by making careful plans at the start of projects and then relentlessly putting them into effect. This chapter describes the careful, painstaking attention to detail throughout projects needed to reinforce partnering and enable project teams using partnering to deliver substantial benefits for everyone involved. Chapter 7, checklist 33 provides a glossary that describes the various kinds of teams used in modern construction.

It is important throughout construction projects that the client's internal team remains fully involved so there is no risk of the project team being distracted by day-to-day issues from concentrating on delivering exactly what the client needs. This is particularly the case when the project aims to produce an original design.

Construction projects move through distinct stages. The sister publication, *Code of Practice for Project Management for Construction and Development*, provides a broad description of the stages in construction projects which is illustrated in Figure 4.1.

This chapter describes the stages that follow those described in Chapter 3 through to completion of the project. At the start of the stages described in this chapter the core team and the lead firm for each technology cluster that has been sufficiently well defined have been appointed. They have worked with the client's internal team to agree the way the project will be tackled. The objectives of these stages are listed in the box on the next page.

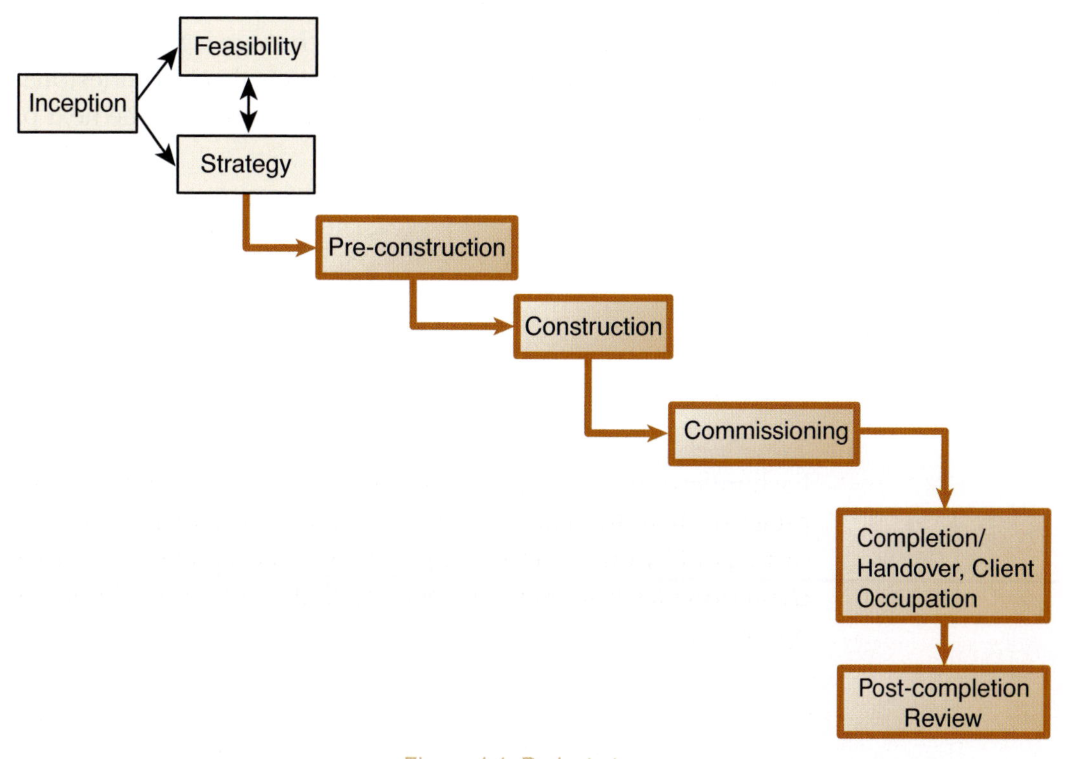

Figure 4.1 Project stages

Objectives of the Main Stages of Construction Projects

Pre-construction	Pre-construction produces a design that can be delivered predictably to achieve the client's objectives, the mutual objectives and performance improvements.
Construction	Construction produces the building or infrastructure in accordance with the design and achieves the client's objectives, the mutual objectives and performance improvements.
Commissioning	Commissioning ensures that engineering systems, mechanical equipment and the installation as a whole has been installed correctly in a safe manner and performs to the requirements of the design.
Completion/handover, client occupation	Completion/handover, client occupation gives the client possession of the completed building or infrastructure, ensures the client knows how the new facility is designed to be run and that there are no defects or that any defects are identified and the client knows how and when they will be rectified.
Post-completion review stages	Post-completion review measures the project's performance, identifies and records lessons for future projects and measures the performance of the new building or infrastructure in use to establish its fitness for purpose and check that it satisfies the client's objectives, the mutual objectives and performance improvements.

The pre-construction and construction stages of most building and infrastructure projects naturally subdivide into shorter stages. These together with the commissioning of the engineering services, handover and project review stages provide a natural structure for construction projects. Each stage involves a distinct technology provided by a supply chain typically comprising many work teams provided by many different firms. It helps to concentrate the project team's attention on their immediate aims if the start of each stage is marked with a milestone (sometimes called a gateway). A typical set of milestone stages for a complex building project is given in the box on the next page.

4.2 Supply Chains

Each milestone stage results from work carried out by one or more supply chains. Most milestone stages require extensive preparation by the supply chains under the overall leadership of the lead firm.

Each milestone stage makes a significant contribution towards project progress. It is the end product of a great deal of preparatory work carried out by one or more supply chains. The lead team of each supply chain provides leadership in ensuring that all the preparation is carried out on time.

Milestone Stages for a Complex Building Project

- Project brief
- Concept design
- Scheme design
- Detail design for each element and system
- Production information for each element and system
- Substructure including underground services
- Structure
- External envelope
- Service cores, risers and main plant
- Finishes and services to entrance and vertical circulation spaces
- Finishes and services to horizontal spaces
- External works
- Engineering services commissioning
- Completion/handover, client occupation.

The nature and timescales involved in the preparatory actions for milestone stages vary considerably. Some involve complex detail designs, research and development, prototypes, specialized training, new construction technologies, manufacturing, innovative quality control as well as construction on site.

The core team is responsible for ensuring that firms and work teams are appointed early enough so that at each milestone all the necessary preparation is complete. The lead time for these appointments may be influenced by long delivery times which therefore need to be identified early.

4.3 Planning Construction Projects

Projects are planned in a consistent pattern of days', weeks' and months' work. Plans define what needs to be complete at each milestone stage and set targets for costs, time, quality, safety and other issues important to the project's success.

In planning construction projects, there are distinct advantages in fitting the milestone stages into a consistent time pattern. There are further advantages in using the universal measures of time. That means each day's work is planned to contribute to a significant week's work that leads to a milestone at the end of each month. The precise tempo of the work is influenced by the size and complexity of the project. Large projects may generate milestones at two- or even three-month intervals while small projects may have milestones that relate to more than one stage. This flexibility is justified because there are deep natural and human reasons for the way we measure time and it is rarely sensible to let project stages impose an alien pattern.

Project planning should begin with a review of the overall approach agreed at the first partnering workshop. The review should check that users' needs are taken into account. It should check that decisions take account of total life-cycle costs, environmental impact and possible changes to the client's business in the future and the consequent need for flexibility in the new facility. It should consider the use of prefabrication and standardized components because of their potential to improve quality and cost. The aim is to ensure that the agreed approach is the best way of achieving the mutual objectives and performance improvements. The risks involved in the approach should be identified and various ways of dealing with each considered. The advice on value management and risk management given in Chapter 7, checklists 29 and 30 will help confirm the agreed approach or identify improvements.

Planning is important on partnering projects because the excitement that comes from putting a well-thought-out plan into effect is remarkably effective in reinforcing cooperative teamwork. This is particularly the case when the team aims at and achieves tough, challenging targets. Indeed it is good practice to consider periodically whether the project team's targets can safely be changed to require even higher levels of performance.

Effective plans start with the agreed mutual objectives and work back to the present time, identifying the sets of actions needed to achieve the objectives. So cost planning begins with the budget justified by the client's business case and allows for every firm involved to be paid a fair and agreed profit and contribution to fixed overhead costs. The money remaining is allocated to provide a cost target for each element and system of the new facility. No contingency allowances are allowed as they merely allow project teams to relax their search for answers that fully meet the client's objectives. More detailed guidance is given in Chapter 7, checklist 26.

Similarly time is planned, starting with the agreed completion date, which should be treated as absolutely fixed. The actions needed to achieve it are defined in milestone stages, which should also be treated as absolutely fixed. Work is then directed to fully completing the planned stage before the milestone is reached. Ideally this will be achieved by work teams during normal working hours because adequate resources have been provided. This does not always happen and teams may have to work extra hours or extra days. In extreme cases an additional team may have to be brought into the project to meet a milestone. This is preferable to allowing a milestone to be missed with the consequential disruption to subsequent stages. Relentless pressure by project teams to make sure they never finish a project late by ensuring they never miss a milestone makes a major contribution to construction efficiency. More detailed guidance is given in Chapter 7, checklist 25.

Quality control should also relentlessly aim at ensuring that work is done properly all the time. This can be reinforced by basing feedback on the number of inspections where no defects are found. If any errors or defects occur, they should be investigated and put right. The first time a specific error or defect occurs may be just a mistake, but if it recurs the investigation needs to find out why and decide whether there is a systematic problem. The aim must be to make sure that the error or defect does not happen again. The investigation should never be used to allocate blame because that distracts the team from their vital task of finding a complete answer. Given this positive approach, quality control can contribute to construction efficiency by avoiding mistakes in design information, making rework unnecessary and instilling a pride in work well done. More detailed guidance is given in Chapter 7, checklist 24.

Safety must be taken absolutely seriously and everyone on site should feel responsible for ensuring that there are no accidents. Unless there is a totally uncompromising focus on safe construction, it is unlikely that any of the project's targets will be achieved. Managers at all levels should take every opportunity to emphasize that safety is fundamental to producing good-quality work efficiently and on time.

An important characteristic of best practice partnering is that it concentrates on controlling time, quality and safety because if they are rigorously controlled, cost control is automatic. It is built into the plans. This is in stark contrast to traditional practice, which too often concentrates on cost control at the expense of time, quality and safety. If work teams are allowed to fail in any aspect of performance, all the targets are put at risk. This is why partnering is tough, hard work. It requires controlled performance in every aspect of work. It requires project teams to concentrate on ensuring that work teams are competent and have all the information they need to stick to well-thought-out plans and supports them by consistently aiming to meet every target.

As each stage of a project is planned, the project team should explicitly identify all the decisions that should involve the client's internal team. The arrangements for these should be agreed and in addition the project team should ensure that the client's internal team understands what is happening during the stage so they can decide what other involvement they want.

4.4 Appointing Work Teams

The core team ensures that competent teams able to use partnering are appointed on time. They should be supported by contracts and project insurance that create a financial and contractual framework that empowers them to do their best work.

Once the approach to be used for a milestone stage is agreed, the core team should check whether new consultants or contractors need to be appointed to provide all the required work teams. The core team should make sure that work teams are appointed sufficiently early for them to contribute fully to the planning of their stage. Work teams usually need to be appointed very early for activities involving extensive preparation, major manufactured or prefabricated components, or systematic information-gathering throughout the project.

Equally the core team should guard against the tendency in construction to focus on interesting issues in too much detail too early with the risk that time and resources are wasted and important aspects are ignored. The core team needs to establish with absolute clarity what needs to be decided at what points in time and who needs to be involved. The aims when a milestone is reached should be that all the preparation is complete, it has been achieved efficiently, and the work teams understand their work.

In making new appointments or reviewing arrangements already in place, the core team must ensure that all the consultants, contractors and specialists are supported by appropriate contracts and project insurance. They should ensure they are acceptable to the client and other key stakeholders. These important issues are described in Chapter 2.

Ideally work teams are part of integrated supply chains of firms in which partnering is well established. In some situations this is possible but in many cases construction projects are faced with less than the ideal. The minimum aim should be that people from the lead firm in each supply chain and work teams centrally involved in design or construction work should use partnering.

4.5 Planning a Milestone Stage

The essential preparation for each milestone stage needs to be planned and carried out in time to allow the stage to be completed efficiently and on time. This is best achieved by treating design, planning, construction and completion as an integrated system.

In advance of each milestone, the preparation for the work involved in the milestone stage is planned and carried out. The preparation is undertaken by the lead team of the supply chain in cooperation with all the work teams that form the supply chain and the project core team. It needs to be done sufficiently early to ensure that the stage is not delayed by inadequate preparation. It is particularly important to ensure that the client's internal team knows that the planning for the milestone stage has begun so they can contribute if they wish.

Planning means that all the necessary work right through to the completion of the milestone stage is identified and programmed in detail. The aim should be to integrate design, planning, construction, commissioning and completion in a seamless system. Indeed planning should take into account the way the completed facility will be used. This is likely to help the supply chain identify opportunities to eliminate waste and inefficiency and represents best practice irrespective of whether or not the consultants, contractors and specialists involved have ongoing responsibilities for the operation of the facility.

The planning of each stage should begin with a review of the agreed targets for the stage. They should be questioned to ensure that they can be achieved and if possible improved. The current feedback measuring the extent to which the mutual objectives and agreed performance improvements are being achieved should be reviewed. If targets are being met, the new plans should support and reinforce the actions delivering the satisfactory performance. If the team's performance is falling short, the plans should include new actions specifically aimed at delivering the mutual objectives and agreed performance improvements. This may well benefit from a value management study taking account of the approach described in Chapter 7, checklist 29.

Next, the agreed approach for the stage should be considered, and alternatives suggested and discussed. The most promising should be tested against the mutual objectives and agreed performance improvements to decide exactly how the milestone stage will be carried out. The risks involved in the approach should be identified and various ways of dealing with each should be considered. The advice on risk management given in Chapter 7, checklist 30 should help the team to be confident about the agreed approach.

In producing the plan for a milestone stage it often pays to consider ways of speeding up the work to generate enthusiasm for fast, accurate work and if possible find more efficient ways of working.

4.6 Induction

Each work team's involvement begins with an induction course to ensure that they have full information about their work, including how it relates to the rest of the project, and are confident about meeting their targets.

Work teams' involvement should begin with an induction course. The content and structure of typical induction courses are described in Chapter 7, checklist 18.

When work teams are involved in preparatory work prior to their milestone stage, induction courses should ensure that each work team understands how their work fits into the supply chain and overall project. The courses explain the project's mutual objectives including the agreed performance improvements. They explain how decisions are made and the control and feedback systems being used. Induction courses aim to ensure that work teams understand partnering. This means that they need to know why they should cooperate in ensuring the success of the overall project and actively join in decision-making. It means that they understand the benefits of solving problems quickly and of being open and not attempting to allocate blame.

The induction course should identify ideas for building on work teams' strengths and compensating for any weaknesses. The induction course provides an opportunity to identify the need for training. This often includes training in the attitudes, behaviour and methods required by project partnering.

Induction courses aim to do everything needed for work teams to be integrated into the project team as quickly as possible. They aim to ensure that work teams clearly understand how they can contribute fully to the project's success and that this will benefit them and their firm.

4.7 Partnering Workshops

The lead team in each supply chain considers the need for supply chain workshops during the work leading up to their milestone stage. The core team considers the need for a workshop at each milestone stage to bring together all key members of the project team responsible for the next stage to confirm progress, review the plan for the next stage and reinforce partnering.

The general pattern of construction projects is illustrated in Figure 4.2 to show where induction courses and partnering workshops fit in.

As a supply chain prepares for their milestone stage, the lead team should consider the need for supply chain workshops. They will normally do this in consultation with the core team and the client's internal team. A series of supply chain workshops may have benefits where the preparatory work is extensive and spread over many months. Many different work teams may be involved and there is merit in ensuring they understand how their work contributes to the overall project.

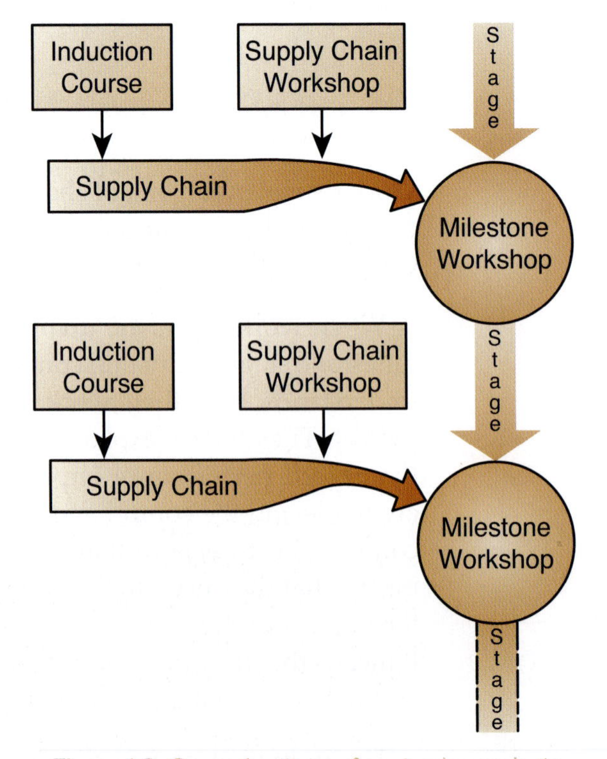

Figure 4.2 General pattern of partnering projects

The key issue in making this decision is the extent to which the work of separate work teams interacts. The more interaction, the more value there is likely to be gained from supply chain workshops. The decision is also likely to be influenced by the work teams' use of partnering. Where partnering is already established, workshops are likely to be an integral part of the supply chain's work. Where partnering is new, more effort will be needed before a workshop can be effective but the potential benefits of improved performance may be greater. This affects the balance of costs and benefits that the lead and core teams have to consider in deciding whether to hold supply chain workshops. The client's internal team will normally be involved in this decision especially when the supply chain aims to produce an original design.

As a supply chain's work reaches the point where it assumes primary responsibility for project progress, the core team in consultation with the client's internal team should consider the need for a milestone workshop. Chapter 7, checklist 32 provide a sample set of questions and issues that can be used in conducting a partnering 'health check' that will help make this decision.

The core team may decide the project is going well and a straightforward milestone meeting is sufficient. Or they may decide that there are serious problems and a separate partnering workshop is needed to refocus the project team on partnering. However, for many projects there is merit in combining a partnering workshop with the formal milestone meeting to create a milestone workshop.

Milestone workshops bring together the whole project team responsible for the next stage of the project. On very large projects this may be a practical impossibility because of the number of people involved. When this is the case, the milestone workshop brings together all the key people in the project team responsible for the next stage. They in turn hold workshops or at least formal meetings with their part of the project organization to ensure that everyone understands their role and its contribution to the overall project.

Milestone workshops have three primary purposes. First, they confirm that the project has completed the current stage so that everything is ready for the next stage to start. Second, they establish a project team confident in using partnering to tackle the next stage. Third, they agree in detail how the next stage will be executed. All this should be checked against the mutual objectives to ensure that these decisions contribute to delivering value for the client and profits for the consultants, contractors and specialists.

Milestone workshops involve the whole project team for the next stage because in addition to new work teams, at least some individual roles change as the project makes progress. Thus each stage is carried out by a team that in some respects is new. This means that there are benefits in discussing and reinforcing actions that support cooperative teamwork even for established members of the project team.

The time needed for the workshop depends on the experience of the project team and the nature of the building or infrastructure. A standard building or infrastructure produced by an integrated team experienced in partnering will normally hold short meetings that include few aspects of a partnering workshop. Essentially the meeting confirms that the project is going well, takes account of any lessons from the previous stage and confirms the team's well-established approach to the next stage.

An original design for a large and complex building or infrastructure produced by a team new to project partnering should have a two-day workshop at every milestone. These should reinforce partnering attitudes and behaviour as well as driving the project forward towards the agreed mutual objectives and performance improvements. Milestone workshops may be shorter in the later stages if the project is going well because there will be fewer decisions to make and less need for team building. In general the core team should maintain a regular pattern of two-day workshops until it is clear that the project will progress equally well if the milestone workshop takes less time. The guiding principle should be that sufficient time is provided for the issues described in Sections 4.8 to 4.15 to be reviewed to ensure that all the essential preparation is complete. Also the guidance given in Chapter 7, checklist 19 applies generally to all the workshops.

4.8 Communication

Partnering workshops aim to foster open and effective communication. The core team monitors communication throughout their project and takes actions to improve flows of information.

Open communication is fundamental to partnering. All relevant information should be shared and nothing hidden or held back. This is essential in making decisions and resolving problems in ways that empower work teams to deliver excellent performance. Anyone found holding back information in the hope of individual advantage should be warned that such behaviour is unacceptable. It may be appropriate for them to be subjected to routine monitoring or audit until they fully accept the benefits of open information.

Appropriate communication methods are fundamental to partnering. The core team should routinely check that there is constant and effective two-way communication throughout the project. In doing this they should remember the main characteristics of good communication. Good communicators give time and attention to understanding what is really important to other people and taking whatever this turns out to be absolutely seriously. They attend to the little kindnesses and courtesies that show other people's points of view have been understood.

Good communicators are skilled at empathetic listening. This means listening until they understand so well that they can rephrase what the other person has said in a way that reflects their feelings. This takes time and training which often includes role-playing in imagined difficult situations so as to practise listening. Replaying situations in which a person behaved badly can also help develop habits of effective communication. This care and attention to empathetic listening is justified because it provides the only secure basis for effective communication, which is an essential element in building teams. Once a person is understood, affirmed, validated and really appreciated, they will be willing to work as part of a team in a whole-hearted way.

Knowing how to be understood is as important as understanding. This means taking account of the other person's concerns and describing your own needs and wants in their terms. Effective people work at deeply understanding other people in this way in order to open the door to creative solutions. Different points of view cease to be a stumbling block; instead they help people find better answers.

Construction projects touch many parts of organizations and communications need to be tailored to the appropriate levels and disciplines so that everyone involved has the information they need. Modern information systems make a massive contribution to efficient communications throughout projects but they can be misused. The core team should ensure that these systems are being used to give everyone the information they need, at the time they need it, and in the form they need it. It is just as important for the core team to ensure that information systems are not swamping people with a mass of communications they do not need. It is all too easy to copy everything to everyone connected with a project. Dealing with marginally interesting information is massively wasteful of people's time. Chapter 7, checklist 14 provides guidance on the use of modern information systems in partnering and Chapter 7, checklist 21 gives guidance on fostering links between work teams.

4.9 Trust

Partnering depends on work teams being able to trust that planned work will be carried out properly. Any persistent problems that threaten this minimum level of trust are discussed and resolved at partnering workshops.

Trust is often described as an essential feature of partnering. This makes little sense because if partnering required people to suspend their normal caution in dealing with others, it would rarely be used. Partnering requires people to take specific actions whatever their motives for taking those actions.

Partnering requires people to behave reliably so that when they say they will take specific actions, other people know they will take those actions. This kind of trust is an essential feature of partnering. It is obviously better if people are motivated by a sense of honesty and good faith, which are the main aspects of dictionary definitions of trust. However, human motivations are complex and difficult to understand. The work of any individual work team on one construction project usually lasts for a short period of time. There is not time to develop a reliable sense of other people's motives. Problems arise when actions by one work team cause problems for others. It is important that any such situations are discussed openly so that all the implications are understood and initial resentment does not fester into adversarial attitudes. Working together to resolve such situations guards against grudges and personality clashes developing. Partnering requires reliable actions and predictable behaviour. This is the essential minimum level of trust that core teams should ensure exists.

4.10 Cooperative Teamwork

Partnering workshops aim to reinforce cooperative teamwork. Partnering depends on work teams taking account of each other's interests. Any persistent problems that threaten cooperation are discussed and resolved at the workshops.

Cooperative teamwork is fundamental to partnering. Attention should be given in all discussions to ensuring that people behave cooperatively. The core team should watch for signs of confrontational or adversarial attitudes and anyone attempting to allocate blame. They should deal with anyone acting against the partnering approach. This is an area where training, coaching and mentoring are of particular value.

It important for the client's representatives to give a lead in encouraging cooperative teamwork based on open and honest discussion of any issues that concern anyone at the workshop. This means reacting positively to any actual or implicit criticism of their own actions so that problems are identified and can be resolved early.

In addition to these ongoing actions, partnering workshops should reinforce cooperative teamwork. This may mean the partnering facilitator encouraging the work teams to look at difficult aspects of their work from different points of view and think about various ways of tackling them. It may be helped by discussing potential problems and risks and challenging the work teams to see these as opportunities to aim at extraordinary performance. The aim is to foster a sense of excitement and confidence about challenging work so the teams are fully committed to their joint success.

Team dynamics change when new people join and new activities begin. It is therefore sensible for the core team to check if any parts of the project team need help in making progress through the distinct stages illustrated in Figure 4.3.

The diagram shows that initially people are wary of each other and have little understanding of each other's views. Given well-run workshops, this initial wariness gives way to a more open approach where people begin to discuss personal issues and show more concern for the views and problems of others. This provides the basis for a next stage in which the team openly discusses how it is working. Purposes are clarified and objectives agreed, information is collected systematically, options are considered, decisions are reviewed and the team works to detailed plans that are jointly agreed. Finally, mature teams work

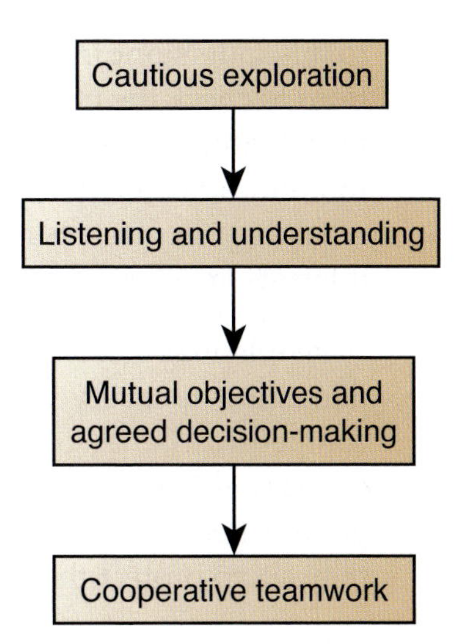

Figure 4.3 Stages in the development of effective teams

flexibly, guided by the needs of their current tasks, and allow the leadership to be determined by the work in hand.

The core team should consider whether work teams are cooperating effectively and decide with the workshop facilitator if special actions are needed at a partnering workshop.

It may be that the workshop should reinforce the idea that joint decision-making is a powerful way of finding the best answers. Examples of innovative designs or construction methods devised during earlier stages of the project could be described. It may need to be explained that joint decision-making is not just a search for consensus. It should be a process of getting various points of view and different ideas in a wide search for the best answer.

It may be that the workshop should remind people of the need to clarify their expectations and bring them out into the open, especially in new situations. It is important that they make their intentions clear and consistently reflect them in their actions to build confidence in their integrity. It may need to be explained that being subtle or clever or using complicated reasoning is likely to be seen as devious or arbitrary.

The workshop may need to include games that illustrate how cooperation depends on people understanding each other. It often needs to be explained that solutions to problems should take account of other people's perception of their current needs. This means getting them to describe their own interests because whatever need they perceive as most pressing, is what will motivate them to accept solutions to problems.

It may be necessary to explain that effective people describe their own feelings and impressions without allocating blame. They encourage others to describe how they feel and why they view the problem that they are discussing in the way they do. They create opportunities for others to describe anything that is bothering them, so there are no hidden problems or concerns. They do not talk about the other parties' attitudes or motives. They do not fall into the trap of attributing their own motives and concerns to others because this nearly always leads to misunderstandings. They listen more than they talk. They ask questions to check the meaning of what the other party has said. They summarize and restate the other party's interests in a positive light in order to build agreement. They check that the other party has understood what they themselves have said. Finally they make sure that their own interests are clearly stated and understood.

People may need to be reminded that competition inside a team is damaging to everyone's interests. They must not passively accept competitive behaviour by other people because that would merely encourage them to continue. Effective people draw attention to competitive behaviour and point out that it is not acceptable. If it persists, they act to protect their own interests.

It is often sensible to remind people that the way they talk about others who are not present shows how far they can be trusted with confidences. If someone uses information given to them by others as a basis for criticism, other people will not be open. This is why integrity is crucial to successful relationships. It means keeping commitments, not making promises that you cannot keep, and apologizing sincerely when you behave badly.

These cooperative behaviours are most effective when they operate within an overall decision-making framework in which people first define the problem to be tackled and then work together to find an answer that takes account of all their interests. When a solution is found, effective people frequently allow others to take the credit for having suggested it. This helps them feel they own it so they are more committed to making it work. Partnering workshops should encourage these effective ways of working.

4.11 Mutual Objectives and Performance Improvements

Partnering workshops check that plans and progress are consistent with achieving agreed mutual objectives and performance improvements. Workshops deal with persistent problems and actively search for more efficient and effective ways of working.

At the start of each supply chain's work, the core team should discuss the mutual objectives and performance improvements with the supply chain's lead firm to ensure that they are being interpreted in ways that are consistent with the client's overall objectives including the agreed budget and completion date. Then supply chain workshops should check that the specific technical solution remains consistent with the mutual objectives and performance improvements.

In addition, all partnering workshops should emphasize that partnering aims to improve performance. It should make clear that this requires work teams to accept that they are responsible for efficiency and effectiveness. No one should passively accept inefficient or ineffective ways of working. They should raise any doubts they have and be willing to question methods and procedures. The guiding principle should be that there is no room in partnering for methods and actions that add no value. The workshop should discuss case studies of how these challenging actions identify and eliminate waste in whatever form it takes. These may include getting rid of time-control procedures that interrupt efficient work, eliminating the checking, double-checking and auditing of costs and making quality control less bureaucratic.

An absolutely crucial feature of successful projects is that a good answer for the next stage exists when each milestone is reached. This should be sought well before the milestone is reached. A key part of this is for each supply chain to identify the best existing answer and check that it will work. This should be done in addition to searching for a better and more innovative answer. Then if no better answer is found in time, the good existing answer is used so the project is not delayed.

4.12 Decision-Making

Partnering workshops check that project meetings provide an effective way of making decisions. Workshops deal with persistent problems and actively foster effective decision-making.

At the start of each supply chain's work, the core team should check with the lead team that the specific technical solution being used is consistent with the agreed pattern of meetings. Advice on effective patterns of meetings is given in Chapter 7, checklists 12 and 13.

All partnering workshops should consider whether the agreed pattern of meetings is still appropriate and that the client's internal team is properly involved. This discussion is likely to be particularly helpful to new people who need to understand how their work fits into the overall project so that they can contribute fully to any meetings they need to attend. Workshops should remind people that meetings bring together all those concerned with a major issue to share information, discuss their plans, review progress, deal with current issues and problems and suggest better ways of working. These focused meetings can be a rich source of ideas that drive work forward. The discussion should also emphasize that partnering means that decisions are based on an open exchange of information and a willingness to explore other people's ideas to find answers that benefit everyone and do not compromise anyone's interests.

Everyone who works in this way should be rewarded by a no-blame culture. People make mistakes and provided they are clearly doing their best in the interests of the supply chain and the project and learn from things that go wrong, it is counterproductive to allocate blame. Indeed a few mistakes can be a healthy sign that people are using their initiative to try better ways of working. Criticism or blame is likely to stifle new ideas and lead to good ideas being missed. People should be encouraged to be realistic about where they can and cannot try new things and provided they act responsibly, mistakes should not be seen as being negligent or reckless.

In guiding workshop discussions, the facilitator should keep in mind established principles of good decision-making and remind participants of them as they work through the four stages illustrated in Figure 4.4.

The first necessary stage is to define the question to be answered. This should be stated as clearly as possible, including identifying the criteria that satisfactory answers need to meet. These should take account of the mutual objectives and agreed performance improvements.

Second, a number of possible answers should be proposed. There are many techniques that can be used including highly creative design methods, brainstorming and similar techniques designed to free groups from thinking of existing answers and help them find new ideas. The aim of the second stage of problem-solving is to identify a number of possible solutions.

Third, the two or three most promising answers are evaluated. They should each be judged against the criteria identified during the first stage. The strengths and weaknesses of each should be listed, particularly taking account of the consequences on other elements of the work. It is important to be seen to be fair at this stage. So, for example, if a variable has to be valued, this is best done by reference to market value, a previous similar case, judgment by an independent professional, normal practice, or some similar objective criteria to provide a fair standard or a fair procedure.

Finally, a decision should be made. This may be to select one of the answers reviewed at stage three. It may be to adopt a combination of elements drawn from several of the potential answers. It may be a decision to evaluate more of the answers identified at stage two or to search for more potential answers. When none of these options appear attractive, the original question may be reviewed. Alternatively, a task force may be set up to search for an answer. Indeed, in major supply

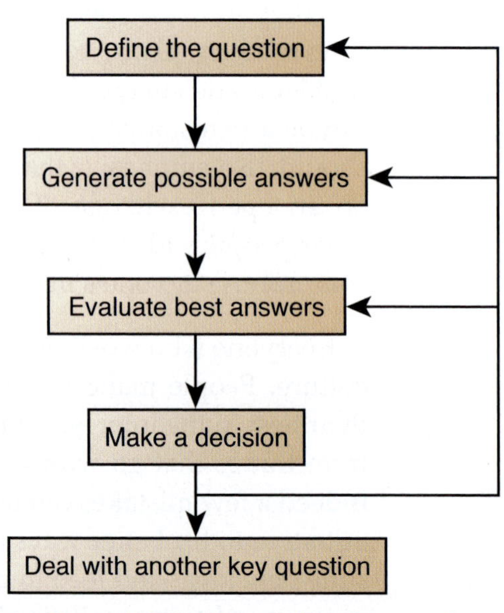

Figure 4.4 Decision-making framework

chains it is common in the early stages to set the same problem for two or three task forces to work on simultaneously. Answers are then presented by each task force in front of the others so that subsequent discussion can provide a more rounded, deeper understanding of the likely source of a good answer.

Once a decision is made another key question can be tackled. The aim is for teams to concentrate on one question at a time to ensure that properly considered decisions are agreed.

As well as ensuring that work teams understand these principles of good decision-making, the workshop facilitator should emphasize that decisions are best made close to the action. Partnering empowers teams to make decisions about their work in the interests of the whole project. Most issues can be resolved by the people directly involved straight away as long as they are well-motivated to cooperate in driving the project forward. Practical answers can usually be found on the basis of a short discussion of the problem in which work team members challenge and question possible solutions. The outcome may be the team realizing why their present way of working is after all the best available, they may find a better answer or they may want to refer the problem to a formal meeting.

Projects using partnering are driven forward through a pattern of regular meetings where difficult problems can be raised and dealt with. An important part of this pattern is a daily meeting of the work team leaders currently on site. This should be held at a consistent time, usually early afternoon. It reviews the current day's work, deals with problems and confirms the next day's work. Typically the meetings take less than twenty minutes. One great benefit of having this regular meeting is that teams do not waste time trying to work out who should deal with a problem because they know their team leader can raise it at the daily meeting.

Like all the project meetings, the daily progress meetings are characterized by ideas being challenged. This must be welcomed provided of course the challenge is rational. The aim is that good ideas are reinforced and adopted while weak ideas are improved or rejected. Knowing that their ideas have been properly discussed and understanding why they have not all been adopted makes it easier to accept decisions and concentrate on putting them into effect. An important principle of cooperative teamwork is that once a decision has been properly made, everyone concentrates on putting it into effect.

This robust, open approach to decision-making should be used in all the project meetings suggested in Chapter 7, checklists 12 and 13.

4.13 Problem Resolution

Workshops support effective problem-resolution techniques.

Each milestone workshop should ensure that work teams understand the agreed problem resolution process and are confident in using it.

The agreed process should take account of the guidance in Chapter 7, checklist 16.

Ideally there will be few occasions when work teams cannot solve a problem for themselves. When partnering is working well, few problems are referred to the core team and it is very unusual for a problem to be referred to senior managers. The workshop should check the extent to which this ideal is being achieved and ensure that senior managers in all the partnering firms, including particularly the client's organization, are aware of the extent to which problems are being resolved quickly without compromising the mutual objectives.

4.14 Performance Improvement

The core team ensures that work teams are supported and encouraged in searching for ways to improve their performance. Problems and new ideas are analysed and actions taken to put improvements into effect on the current or future projects.

The core team should actively encourage work teams to suggest ways of improving their performance. Suggestions may emerge during routine project meetings or informal discussions. Some projects have found that a suggestions box encourages ideas for better ways of working. A partnering workshop may decide that some aspect of the project could be improved. This may be triggered by a specific problem or a good idea about how the project should be tackled.

Beyond these spontaneous sources of good ideas, the core team should routinely review problems that have arisen and been solved by work teams to check whether they suggest opportunities for improving performance. They should particularly focus on any problems that influence the value delivered to the client or users or that suggest a systematic weakness in the team's approach.

Whatever the source, performance must be defined and measured before it can be improved. This should begin by analysing the processes involved in a problem or idea for performance improvement. The process analysis should identify the main inputs, activities, constraints and outputs. Key factors should be measured to establish the extent and location of the problem. A target should be set for improvement and possible actions identified. These should be discussed widely so the whole project team can contribute to finding a better approach. The process is most likely to produce a good answer if several alternative answers are considered, evaluated and discussed before selecting one to put into effect. Chapter 7, checklist 17 provides detailed guidance on this approach to identifying performance improvements.

At this point the core team must decide whether the new approach can be used immediately or if it requires such fundamental changes to the way the project is being run that it should be documented for use on future projects. The client's internal team should be closely involved in this decision.

When the core team decides to use a new answer, it must be taken into account in the project's plans, procedures and feedback systems. The core team should pay particular attention to the resultant feedback and take further actions if the intended improvement is not achieved.

The development of ideas for improving performance may take place at project meetings or partnering workshops. It is more likely that a task force will be set up to produce an answer quickly. This is often the best way of ensuring that a project is not delayed by key people being distracted by spending time and effort developing an interesting new idea. The use of task forces is described in Chapter 7, checklist 22.

4.15 Feedback

Work teams have feedback at least every week about their own performance. Projects have feedback at least at every milestone about the overall performance. Clients and other stakeholders have regular feedback about progress and decisions that affect them.

Every work team should have feedback on their performance measured against their own targets and the overall supply chain or project targets at least weekly. Work teams should be encouraged to discuss their feedback to see if it provokes ideas for changing, adapting or developing the way they work. Provided the project is progressing in accordance with overall plans the core team review a summary of progress each week.

The project team's overall performance should be measured at each milestone and reviewed by the core team prior to the milestone workshop. The issues that need to be taken into account in designing and using feedback systems are described in Chapter 7, checklist 15.

It is particularly important that the client has feedback so they understand the project team's decisions about the new building or infrastructure. This feedback should take account of the implications for the client, users, neighbours and other stakeholders. It is particularly important to consider the form this feedback should take because many people do not fully understand drawings and can be misled by professional descriptions of the performance and quality standards being provided.

This means the core team should ensure that all aspects of the project likely to affect any of the stakeholders are properly understood. This may mean explaining design concepts and assumptions, producing prototypes or virtual reality representations and organizing visits to the construction site. These actions should include explaining the reasons for not providing things that some of the people involved may have hoped for in the new building or infrastructure. In other words the core team should actively manage the expectations of all the stakeholders. The aim is to ensure that when the client takes possession of the completed facility, there are no surprises.

4.16 Core Team Meetings

The core team reviews progress every week to ensure the project is meeting the client's objectives. The core team drives the project towards its agreed objectives and ensures that the Project Execution Plan and Project Handbook reflect the agreed approach.

The core team should meet every week to ensure that the client's objectives are being achieved. This can usefully be combined with the core team's review of quality, time and cost. Provided the project is on target, the core team concentrates on anticipating problems and looking for ways to further improve the team's performance. This includes considering whether any of the project team's targets can be revised to require higher levels of performance.

Some core teams select one of their members to chair all their meetings. Others rotate this role, taking account of the subjects to be discussed. This has the advantage of allowing different ways of running the meetings to be tried.

The core team's meetings should be provided with up-to-date feedback on project performance. This should compare actual performance with the planned quality, time and cost targets. It should also identify problems that work team's have been unable to solve and opportunities to improve performance on which a decision is needed.

In addition to their own meetings, members of the core team attend key project meetings and partnering workshops. They also regularly walk around the places where project work is underway and ask questions about the project and its targets and progress. The purpose of these activities is to ensure that they have first-hand knowledge about progress to help them understand the feedback reports. Suitable questions for them to ask are given in Chapter 7, checklist 15.

The outcomes of core team meetings should be decisions that support things that are going well, actions to resolve problems and decisions about opportunities to improve performance. The core team should also check that the Project Execution Plan and the Project Handbook are kept up to date.

4.17 Membership of the Core Team

The membership of the core team should be reviewed prior to each milestone to ensure that all the key interests and knowledge are properly represented.

The membership of the core team should be reviewed prior to each milestone stage to ensure that all the interests likely to have a significant impact on the project's success are represented. The majority of the members of the core team remain as originally established throughout projects but it may be sensible for the lead firm in each of the supply chains currently central to progress to be represented on the core team. Some core teams go further and include everyone who could have a significant impact on the current milestone stage. The box below lists various responsibilities and disciplines that may be included in the core team. These decisions should be balanced against the benefits of keeping the core team small and maintaining a consistent membership.

People that may be Included in the Core Team

- Owners
- Company advisors
- Senior managers
- Occupiers
- Users
- Facilities managers
- Maintenance engineers
- Local authority representatives
- Planning experts
- Neighbours
- Local special interest groups
- Lawyers
- Production engineers
- Health and welfare managers
- Safety engineers
- Fire consultants
- Designers
- Construction managers
- General contractors
- Specialist contractors
- Manufacturers
- Procurement managers
- Quality control managers
- Time control managers
- Cost control managers
- Information and communication experts.

4.18 Final Workshop

The final partnering workshop captures lessons for use on future projects and celebrates a successful project.

A final workshop provides the opportunity for lessons from the project to be discussed and recorded for use on future projects. In this way the final workshop plays a crucial role for the consultants, contractors and specialists involved by identifying ways for them to improve their performance. As part of this it often helps to discuss how feedback is used for the benefit of future projects to ensure that all the firms have good feedback and information systems in place.

In preparation for the final workshop, the project should be measured against the original business case, the client's objectives, mutual objectives and agreed performance improvements to establish how successful it has been. This review should pay particular attention to the operation of the new facility and the views of users. It should also consider opportunities for the consultants, contractors and specialists involved to work together in the future. As part of this they should consider the need to develop their approach to partnering further. This may identify new practices and procedures they should adopt. It may even suggest the need for cultural change. The review may also identify specific opportunities for individual development.

At the workshop, the reasons for successes and failures should be discussed and lessons identified. The workshop should particularly look at any aspects of the project that did not work well. These may include designs that caused problems, control systems that failed to provide an early warning of problems or construction techniques that worked badly. The workshop should agree how each lesson will be recorded for the benefit of future projects.

If the core team commissioned a case study of the project's approach and performance, it will provide useful background information for the workshop. Consideration should be given to publishing the case study and whether it needs the assistance of professional journalists and designers. Case studies should be publicized so that more people know what works well in what circumstances. The construction industry press and conferences provide many opportunities to describe successful projects. Even the general media will describe outstanding successes on major projects. They are keener to publicize failures but everyone in the construction industry has an interest in seeing this tempered by descriptions of successes.

The workshop should consider whether any aspects of the project's performance need further work to determine what has been achieved. These may include the influence of the new building or infrastructure on users' effectiveness and efficiency, the running and maintenance costs and the environmental impacts.

It is important that the workshop agrees how all the feedback will be made widely available so it is used on future projects. Individuals in the client's organization and all the main consultants, contractors and specialists involved should be charged with reporting to their senior managers at six-monthly intervals on how feedback from the final workshop is being used. It is helpful for this to be discussed by groups of senior managers from all the firms involved. This is an important part of building the long-term relationships between clients, consultants, contractors and specialists that underpin the most effective forms of partnering as described in Chapter 6.

The conclusion of the final workshop provides an ideal opportunity to celebrate a successful project. It is good to mark the end of the project with a party to reward the project team for a job well done. Team members should be awarded prizes or certificates in recognition of their contributions to the project. This all helps cement partnering relationships, strengthens the relationships between the consultants, contractors and specialists involved in the project and helps create an efficient construction industry.

Oxford citizens steel ahead on partnering

Case Study Reference: Housing Forum Innovation Case History

Oxford Citizens' Housing Association had established a successful partnering arrangement with Oxford City Council and contractor Willmott Dixon. This had a good track record with traditionally constructed projects, including Oxford's Blackbird Leys estate.

The partners were keen to reduce project times and defects. They decided to explore the advantages offered by the Corus Sure Build steel frame system, which seemed to require less time on site and produce a more predictable end product. These advantages suited the brief for the £2.7m William Morris Court 36-unit development in the Barton area of the city.

Encouraged by Willmott Dixon's advice, Oxford Citizens' Housing Association decided that a steel frame could radically cut the programme time and improve the finished quality of the development. They also expected the timescale to be more predictable which meant that tenants could occupy the scheme with a minimum of delay.

The primary partnership agreement was between Oxford City Council, Oxford Citizens' Housing Association and Willmott Dixon. However, a wider and more informal partnership drew in other players who helped ensure the new form of construction worked well. One important action was that each of the key organizations involved appointed a 'tzar' to expedite the decision-making process and ensure their united commitment to the project.

Although this was a trial project for an innovative steel frame, the partners kept a firm hand on cost. The final cost was £100,000 under the £2.68m guaranteed maximum price. Oxford Citizens' Housing Association re-invested its 50% share of the saving in improving the end product.

The programme for a comparable masonry building would have been 82 weeks according to Willmott Dixon's business development director Philip Stephens. He reported: 'The programme for the Sure Build system was 64 weeks, yet despite scheduling problems we lost only two weeks. Once the steel erection got going it was largely unaffected by record-breaking wet weather that would have caused big delays with masonry.' This suggests a time reduction of about 25%.

There were very few defects on completion and far fewer than with traditionally constructed projects.

Partnered IT project turns the tide in Dundee

Case Study Reference: M4i 190

An innovative project led by Dundee City Council Architectural Services secured the Best Team award from the Convention of Scottish Local Authorities in 2000. The key to their success is a fundamental change in procurement strategy for installing networked computer suites in Dundee secondary schools. Chief Architectural Services Officer, John Porter, led the team that embraced two of Egan's suggested initiatives: partnering and electronic exchange of information between the partners. 'As the biggest construction client, the public sector must take the lead,' said Porter.

The £2.4m project equipped nine schools to National Grid for Learning standards including access to the World Wide Web. Dundee City Council used partnering with four contractors in a pilot that will lead to the best being selected for ongoing networking services. A guaranteed maximum price was agreed for each school with incentives to achieve a lower target. Cost savings on each installation were used as a basis for setting a tougher guaranteed maximum price for the next, feeding a continuous improvement cycle.

The benefits resulting from partnering included the following:

- The Convention of Scottish Local Authorities award showed the project as performing better than industry norms against a broad range of criteria.
- Cost – the demand for networking services was buoyant. Dundee City Council's Property Services Co-ordinator for Education, Derek Currie, had been facing annual cost increases of 10%, so he could barely conceal his delight that the first phase schools were completed 8% under budget: 'The contractors delivered on average 5% below the guaranteed maximum price and our in-house consultants billed us 30% less than we were expecting.'
- Construction time – disruption to school routines was minimized and all the installations were completed at least 10 days (25%) ahead of schedule. A whirlwind of change seemed to have swept through Architectural Services as well. Chief Architectural Services Officer, John Porter, reported: 'Normally architect's instructions account for between 10 and 30% of delays but none were issued, a testament to the teamwork between the Education Department, the schools, the contractors and us.'
- Predictability – all installations were completed ahead of programme and within budget. It was therefore no surprise that the Key Performance Indicator for customer satisfaction leapt from an expected 7 to 9 out of 10.

Partnering for performance and value

Case Study Reference: CBPP176

In August 1996, client South West Water appointed three partners, contractor Morgan Est Water Division Civil Engineering, designer Mouchel and process manager Degremont, to refurbish a wastewater treatment plant in Plymouth within a tight schedule and budget. The partners worked as an integrated team and, by June 1999, achieved their cost and programme goals, enabling South West Water to meet tough new legal requirements.

On the appointment of the three main partners, a series of workshops was organized, all facilitated by an external consultant.

Milestone workshops for interim progress reviews and planning were held on site at varying intervals during the construction period of 18 months. Participants included site managers and supervisors, such as project manager, subagents, section engineers and key subcontractors' representatives. One inspired move was to invite staff who operated the wastewater treatment plant to the workshops. As end users, they contributed much to the development of the project.

All subcontractors were involved early in the design process, in order to harness their expertise and make them feel part of the team. Specific meetings were scheduled to draw on their specialist inputs and ideas. Key benefits from the partnering approach included:

- More certainty in maintaining programme and budget.
- Improved buildability as a result of close liaison between the client, designer and contractors.
- The opportunity to apply value management principles continuously as a team.
- The outturn cost of civil works was approximately £100,000 below the target cost.
- Successful minimization of waste (e.g. concrete waste reduced to 1.5%).
- Reduced reworking by doing work 'right first time'.
- No claims submitted by construction firms or their subcontractors working under the partnering agreement.

Client opts for partnering in an innovative project

Case Study Reference: Building Services/ Education BS 185

Between 1994 and 2000 the Imperial College of Science, Technology and Medicine in London succeeded in reducing its annual energy costs by 33%. It is estimated that the installation of a combined heat and power system, a £9 million project, delivered 24% of this saving. The project used a partnering arrangement between the College, the main contractor, London Electricity Services, and a number of suppliers. The installation was completed with minimum disruption to the College's academic activities and ensured essential, round-the-clock continuity of energy supplies.

When partnering was being introduced, the contractor partner hosted a one-day workshop that included consultant teams. The event was designed for team-building and for discussion of major issues affecting the project. One key decision was to use the Engineering and Construction Contract for the project.

Immediately following the meeting the partners began planning the sequence of the 32 project packages in order to establish start times on site for everyone involved.

Further workshops were held where partnering and project objectives were top of the agenda. The workshops included team-building exercises and an open forum that allowed participants to raise problems and suggest good ideas.

Several of the people involved were sceptical about partnering at the beginning but were completely on board by the end of the workshops. More importantly, trust among team members, including subcontractors, was built up at the workshops and became an important advantage of partnering.

The 'soft' issues of communication and motivation were the greatest challenges. The first few meetings between the College and construction teams were helpful in setting the framework for dispute resolution, as members learned how to resolve issues by sitting down together and discussing the best way forward. This expectation of cooperative behaviour was formalized by a dispute resolution policy, which stipulated that any dispute must be settled as it arose.

The key benefits that were helped by using partnering included the following:

- The College's energy costs have been reduced by 24%.
- The College has become self-sufficient in heating.
- The College has lower running costs for cooling and air conditioning.
- Increased confidence that the partnering approach works as expected.
- Increased confidence in undertaking work within budget.
- Minimal delays in the construction programme despite the work being carried out in a busy university.

5

Developing partnering skills

5.1 Introduction

Partnering requires consultants, contractors and specialists to organize themselves internally to support individuals and teams in using cooperative teamwork. This internal partnering is necessary to obtain the full benefits of partnering.

Partnering begins by concentrating on external relationships. It requires consultants, contractors and specialists to work in cooperation with others and take account of their interests and concerns much more than is normal in traditional business relationships. Firms make this change because working in cooperation with others helps them achieve high levels of performance and innovation.

However, the external focus causes internal changes, which initially tend to happen in relatively unplanned ways. Once a firm is seriously involved in partnering, a point is reached where the new way of working needs to be supported by internal changes. This means using cooperative teamwork internally to support partnering arrangements. The new way of working is called internal partnering.

Internal partnering means all the internal arrangements needed to encourage and support work teams in cooperative teamwork. More than this it means bringing any internal support services such as finance, accounting, audit, legal, marketing, purchasing and human resources into partnering arrangements. Helping specialists to understand partnering and its benefits and involving them in cooperative teamwork bring several important benefits. At the very least specialists do not feel threatened by the need to become part of cooperative teams and so are unlikely to obstruct the changes that partnering requires. Ideally they will help work teams search for more efficient ways of working.

Internal partnering requires commitment from top management and considerable delegation of authority over decisions and actions. These changes can be difficult in large organizations whether they are clients, consultants or contractors. The variety of specialist roles and responsibilities within traditional organization structures make it difficult to act consistently in working in cooperation with other organizations. Smaller organizations face different barriers to change. Power and authority may be too centralized. People may claim to be too busy to think about change. So most senior managers charged with implementing internal partnering face a tough challenge.

5.2 Internal Partnering Team

Consultants, contractors and specialists should establish a strong internal partnering team that provides leadership in using partnering and making the organizational changes needed to support partnering.

Firms should establish an internal partnering team of senior managers to lead their use of partnering. It usually includes senior managers involved in establishing partnering arrangements, particularly those working in ongoing strategic arrangements.

The internal partnering team's role is to ensure that the organization is able to play a full role in partnering arrangements. It needs to establish a genuine commitment to partnering from top management. It needs to support senior managers in establishing new partnering arrangements and those involved in strategic teams with other organizations. It needs to ensure that work teams are supported in using cooperative teamwork.

It is vital that the internal partnering team is given the necessary support and direction by top management. It should not be left to operate alone, nor should it be restricted by outdated controls. Partnering is unlikely to flourish if, for example, the internal partnering team is expected to perform within a rigid budget determined traditionally on the basis of market prices rather than total value. This is not to suggest that financial considerations can be ignored. Indeed it is vital that costs and benefits are fully considered before changes are made. To demonstrate this, the internal team should produce a business case that explains their approach to internal partnering.

5.3 The Business Case

Internal partnering usually begins with a business case in financial and other benefit terms. This should as far as possible fit the firm's existing strategy. It should set clear, achievable targets that start modestly. Frequent small steps are better than great leaps.

The internal partnering team should arrange for a business case to be prepared based on wide consultation throughout all levels of the firm. It is also sensible to bring suppliers and customers into early discussions, either singly or as a group. Some companies use open days to explain partnering and answer questions. These sessions help the internal partnering team to assess the interest and commitment amongst key partners and potential partners.

The business case should provide a description of the firm in five years' time if it fully adopts partnering in all its external relationships. It often helps to compare this to several alternative scenarios so the costs and benefits of partnering are clearly stated. The business case should describe the changes needed to realize the full partnering model of the firm. This should be seen as a series of distinct stages. Overall this plan should aim at ambitious objectives but should not expect too much too quickly. It is often the case that small, frequent steps are preferable to giant leaps. The following should be taken into account in setting targets:

- The firm's strengths and weaknesses.
- The competence of the firm's personnel and their success in working together and with outsiders.
- The demand for construction and the nature of the local construction industry.

- The attitude of clients, consultants, contractors and specialists to partnering.

The business case should emphasize that internal partnering requires considerable effort, especially in the formative months. It is tough for people to stop using an adversarial approach in which they hide problems, evade responsibility and blame others. It takes courage to expose problems, accept joint responsibility for solving them and let others take the credit for good ideas. The fundamental nature of these changes make it inevitable that there will be at least some opposition at all levels so the business case needs to be robust. It should take account of the essential ingredients of successful partnering relationships including the following:

- Genuine commitment from top management to the use of partnering in the firm's external relationships.
- Clear communication of the objectives and long-term goals throughout the organization.
- A partnering ethos that ensures partners will not act against each other's interests.
- Open financial arrangements that share rewards in a manner perceived to be fair.
- Sufficient resources, especially in terms of the quality and number of competent specialists.
- Full, candid, frequent and open communication and exchange of information internally and externally including open-book accounting.
- Procedures to support the discussion, understanding and addressing of partners' needs, expectations and problems.
- Procedures and systems integrated with those used by partners.
- Procedures to encourage innovative ideas and solutions including working with research and development specialists and other organizations able to provide original ideas.
- Effective mechanisms for resolving problems at the appropriate level.
- Agreed, measurable and realistic performance indicators and effective feedback systems.
- A basis for establishing best practice and then improving it year on year.

The business case must be supported by a detailed and rigorous financial analysis that takes into account all the costs and benefits. Partnering makes consultants, contractors and specialists more competitive, so benefits can be measured in terms of increased profit. It is usual for the financial benefits to become evident early in partnering arrangements. However, the full benefits are more likely to emerge in the medium and long term. Like all processes of continuous improvement, rewards accumulate progressively.

The business case should not rely only on financial figures. Doing so seriously understates the full benefits and can produce weak or even wrong conclusions. Partnering provides many benefits that cannot be measured exactly and so the business case inevitably includes some qualitative measurements. Value and risk management studies may help ensure that all the benefits are taken into account. Chapter 7, checklists 29 and 30 provide relevant guidance.

The changes proposed in the business case to put internal partnering into effect must fit the firm's overall strategic plan. This may mean changing the overall strategy but whatever is needed, the two must complement each other.

5.4 Maintaining Change

Once the business case is accepted the internal partnering team should set up mechanisms for driving partnering into all the parts of the firm.

The internal partnering team should meet regularly to monitor progress, measure performance improvements and deal with problems. The frequency of the meetings should be determined by the stage the firm has reached in using partnering. In the early stages, the meetings will be frequent and be closely linked to the day-to-day activities of staff involved in partnering arrangements. As partnering becomes established, the meetings will become less frequent and will concentrate on longer-term, strategic issues.

In leading these changes, the internal partnering team should be a source of clear information about partnering and partners for fellow employees. The firm should already hold detailed information on major customers and suppliers but the move towards partnering may require information about areas the firm has not previously considered. Information about potential partners must be interpreted carefully. The potential of a small, young company might not be immediately apparent from the answers in a survey. A supplier that is currently no more than competent might improve with advice or help from new technology. The internal partnering team should ensure that the firm's marketing information is designed to identify companies who share their ideas on partnering.

As new partnering relationships are considered, the internal team should make sure that partners are chosen with care. They should review the extent of new partnering arrangements and consider their aims, timetable and procedures. The internal partnering team should concentrate on principles and not deal with all the specific, detailed and often technical issues raised by individual partnering arrangements. It often makes sense to set out a framework of principles that teams throughout the firm should take into account in forming partnering agreements.

There is no universally correct sequence of actions for an internal partnering team. It may be most effective for them to begin by giving the firm's work teams already involved in partnering full decision-making powers about their own work. This often involves substantial changes, including bringing specific expertise into work teams, establishing effective communication links, establishing measurable performance indicators, and investing in training and coaching.

It may be more sensible to concentrate initially on supporting senior managers involved in strategic arrangements. They need the authority to take key decisions about their organization's role in the confidence that the rest of the firm will act on the basis of those decisions.

As the internal partnering team takes their first initiatives, the need for other urgent changes to support the new way of working will become apparent. The change from an organization based on managerial instructions to one designed to support empowered work teams will affect everyone in the firm. It is the internal partnering team's responsibility to identify and relentlessly drive these consequential changes into all parts of the organization.

In making plans, the internal team needs to accept that partnering takes time – years, not months – to develop properly. It is sensible to set some targets that can be achieved in a short time period to give some quick successes. It is also sensible to realize that targets requiring major changes or developments may take years to achieve. It should be an ongoing part of the internal team's work to measure progress, solve problems as they arise, fine-tune objectives and provide encouragement.

In ensuring that the firm plays its full part in partnering with other organizations it helps to acknowledge that the firm's processes and systems are capable of being improved. It also helps to recognize that partners may have much to teach the firm. The internal partnering team should be open to ideas from any source but make sure that any changes have substantial net benefits.

As changes are put into effect, the internal partnering team should keep in mind that problems are frequently encountered when partner firms are managed differently. They should try to reduce differences in business methods, objectives, and strategic ambitions between their own firm and key partners. This may mean that internal changes need to be coordinated with changes in partner organizations. Working with partners in this way may identify areas where they can all benefit from mutual help. This can include helping partners with advice, financial assistance, exchange of staff, joint training, access to specialized resources, etc.

The internal partnering team should also keep in mind that partnering arrangements have life cycles. They emerge, develop, reach maturity, face crises and change or decline. Many partnering arrangements go through the following life cycle several times:

- Assessment of potential partners' skills, goals, efficiency and financial strengths; the tasks and processes involved in possible joint activities; and the environment in which they will be pursued.
- Negotiations that combine formal bargaining and less formal activities aimed at understanding each other.
- Commitment that includes formal legal contracts and a psychological contract.
- Execution of the joint work involving many personal interactions.
- Re-evaluation and revision of the partnering arrangement.

The internal partnering team needs to be clear where they are within this overall pattern in each of the firm's partnering arrangements and make sensible preparations for the next stage.

The internal partnering team should regularly determine how effective they are being. This judgment should take account of formal feedback from partnering arrangements. This should be supplemented by individual members of the team walking round the places where partnering projects are underway and asking questions of the people involved. The box below provides suitable questions.

Questions for People Involved in Partnering Arrangements

- What are the agreed mutual objectives of your partnering arrangement?
- What are you doing to help ensure they are achieved?
- What is your role in quality control?
- What are you doing to help ensure there will be zero defects on completion?
- How do you contribute to safety management?
- What are you doing to help ensure a safe project?
- What is the agreed completion date?
- What are you doing to help ensure it is achieved?
- What is the agreed budget for the project?
- What are you doing to help ensure it is achieved?
- Is the firm meeting all its quality, time and cost targets?
- What are your immediate objectives for this week?
- What may prevent you achieving your objectives?
- What could the firm do to make a significant improvement to your performance?
- Do you have all the information you need to do your work to the best of your ability?
- Are the agreed decision-making procedures being used?
- Are problems resolved quickly?
- Do you have any problems that have not yet been resolved?
- What significant improvement to normal performance is being achieved on this project?
- What is your contribution to ensuring this is achieved?
- Are you being well paid and is the firm making a good profit?

Any concerns raised by the answers should be dealt with quickly in discussion with the immediately responsible people. Serious problems or smaller problems that suggest a growing trend should be discussed at the next internal partnering team meeting. Problems should be solved in ways that make it unlikely they will recur. However, in solving immediate problems, the internal partnering team must keep in mind their main responsibility of ensuring that internal systems and procedures support partnering.

5.5 Fundamental Changes

Cooperative teamwork requires people to be open to new possibilities, new alternatives and new options. Cooperative teams value differences, build on strengths and compensate for weaknesses. This means fundamental changes for people used to traditional practice.

Partnering requires a substantial change of attitude for many people. Many people have grown up in the belief that individual companies must compete. They must concentrate on looking after their own interests. The way to stay in business and make profits is to beat other companies. These traditional beliefs breed adversarial attitudes which are deeply ingrained in many parts of the construction industry. The most effective way for the internal partnering team to help people adopt a cooperative attitude is to focus on success. This means encouraging discussions throughout the firm about what is needed for construction projects to be successful for everyone involved. When there is some agreement on what needs to be done, the discussion can be widened to identify and agree the most effective way of making the necessary changes.

It is not easy for an entire company, however small, to make a full commitment to cooperative working. 'Them and us' attitudes have to be identified and discussed so they can be replaced by agreed objectives and cooperative ways of working. This involves:

- Understanding that all customer and supplier relationships, internally and externally, benefit from using partnering.
- Developing a new culture that supports cooperative teamwork, open communication and win–win attitudes.
- Encouraging open and honest communications, internally and externally.
- Welcoming changes that improve performance.
- Giving and receiving feedback.

Cooperation is gradually being understood and its benefits recognized. People recognize that through work and social interactions some relationships grow so close that they become highly interdependent. The people involved trust each other to behave in specific ways so each individual does not have to develop all the skills and collect all the information needed for joint activities because it is available through their relationship. This makes them stronger, wiser and able to achieve more. People come to realize that as they help their partners become more successful, they benefit themselves. Eventually, mature people grow to realize that all their interdependencies are interconnected. They achieve most when they are all successful and they lose most when they all fail.

Leading practice in the construction industry now recognizes that self-centred, independent behaviour is unsuccessful. It has established some effective partnering arrangements and is now extending these fledgling interdependencies into long-term, strategic arrangements. These have the potential to create a hugely rewarding environment, offering the opportunity of success for everyone involved. It requires the sustained and committed effort of clients, consultants, contractors and specialists over a significant period of time to reach this world-class performance.

The task is daunting because, like many other major industries, construction faces an ever-accelerating rate of change. This has caused society in general and the construction industry in particular to fragment into narrow specialisms. Modern life is so complex that it is impossible for any individual to be able to see the whole picture involved in any modern industrial activity. It is hard to keep track of the people they interact with directly, let alone those that do not have an immediate impact on their work. Leading consultants, contractors and specialists increasingly recognize the need to constantly review and change their structures and policies to support partnering. They understand that this is essential for their own long-term survival.

5.6 Company Structure and Policy

Many consultants, contractors and specialists need to change their structure and policies to fully support partnering. Managers need to be involved in more external than internal communications. Work teams need the authority to make decisions and take actions as part of project teams. Management hierarchies are downsized. Policies are flexible to fit the needs of individual projects.

The internal changes needed to use partnering effectively are not easy. Achieving them requires individual firms to understand where they fit into a complex pattern of relationships formed by project teams and supply chains. The full benefits of partnering depend on cooperative teamwork being used throughout project teams and supply chains. Consultants, contractors and specialists should identify their own key external relationships and then shape their internal organization to support them.

The first and most obvious requirement in planning internal partnering is for consultants, contractors and specialists to operate consistently in dealing with other organizations. This requirement makes it essential that internal departments cooperate. The principles described in Chapter 7, checklist 23 will help the internal partnering team lead the necessary changes to the firm's structure and policies.

The internal partnering team needs to help their firm develop into a self-organizing network that works and thinks long term. They need to make decisions on the assumption of reciprocity in indefinite, sequential transactions. Obligations are often implicit rather than explicit. Contributions from each party are balanced over the entire exchange relationship and are not expected to be equivalent in each and every transaction as is the case in market transactions. In markets, the standard strategy is to drive the hardest bargain in each exchange. In self-organizing networks, the aim is to create indebtedness and reliance over the long term.

Effective networks enable organizations with complementary skills and knowledge to learn from each other. For useful learning to take place partners need to see the world sufficiently differently to enable each to be exposed to knowledge they would not have captured for themselves but be sufficiently close in cognition and language to allow meaningful communication.

This level of interdependence evolves slowly, beginning with minor transactions in which little risk is involved and both parties can prove their trustworthiness. Success at this low level helps them to expand relationships and move on to major transactions. Interdependence becomes an integral part of the relationship. Problems are resolved within the relationship and gradually a mutual orientation emerges. This is expressed in a common language to discuss technical matters, contracting rules, and standard processes and products. In time this grows to deal with business ethics, technical philosophy, and the handling of problems. The resulting well-developed mutual orientation provides a set of more or less explicit rules that limits opportunistic behaviour and so saves the costs of forming and using contracts.

The most effective forms of interdependence provide a kind of loose coupling which preserves some autonomy for partners. Consultants, contractors and specialists respond to changed circumstances and benefit from a more or less stable framework of cooperative teamwork. This reduces the risk of cumulative misjudgments and misdirected learning by exposing people to alternative points of view. It allows networks to access various sources of information and provides for the interactive learning needed in high-technology innovation. Modern organizations have to accept some ambiguity in the perceptions and orientations of individuals because this provides a greater number of potential solutions than are available in rigidly controlled firms. Loose coupling gives self-organizing networks the capacity to respond to changes in their environment. They are more flexible than firms organized on a traditional basis.

Partnering firms invest in education and training to enable everyone to achieve their full potential. They use modern apprenticeship schemes to ensure that their workers have the right mix of skills. They invest heavily in research, development and design, typically at least double the percentage of turnover of more traditional competitors.

This increasingly involves developing new technologies and then training workers to apply them flexibly in response to new situations and rapid change. As a result, partnering firms are innovative and productive because their committed workers have high skills and are supported by high levels of investment. This is reinforced by basing reward systems on long-term performance. The salary gap between the highest and lowest paid staff is much smaller than in many large Western companies where directors' pay may be several hundred times that of their basic workforce.

The same deep acceptance of cooperative attitudes is often reflected in partnering firms accepting the need for trade unions to represent the interests of their staff. They work with the unions to provide generous pensions and health care for their staff and their families. People are treated generously when they are sick, injured, unemployed or otherwise disadvantaged. When change forces partnering firms to make workers redundant, they help them find new employment including providing training and finance. In return, trade unions act responsibly by helping firms develop and grow because they know this is the best way to ensure the long-term future of their members. Trade unions and senior managers work in cooperation to find the best answers to problems. Answers take account of the needs of all the stakeholders.

Partnering firms have family-friendly policies that typically include generous maternity and paternity leave. They provide good childcare for working mothers.

Partnering firms develop close ties with the local communities that provide their workforces. Partnering firms understand the importance and value of the public realm and actively engage in local issues to foster a healthy and vibrant community.

Partnering firms develop close ties with customers, suppliers and subcontractors. They bring workers into their decision-making processes. They understand and act on the reality of their mutual interdependence. Risks are collectively shared.

A number of factors have led to the increased use of networks. There has been a speeding up in the rate of innovation which has overextended the scope and capability of single organizations. Increasingly new products bring together several technologies, mastery of which is owned by separate consultants or contractors. Indeed important markets increasingly require the integration of a variety of new products and services into new systems. Faced with these developments it is important to keep in touch with new developments, and networks provide all the consultants, contractors and specialists involved with a means of doing this at a low cost.

Relations within networks consist primarily of agreements to share research, and to provide manufacturing and marketing and finance. Networks result from relatively stable and recurring patterns of such interactions. However, some developments involve choices that lead to the strengthening of some links and may lead to a weakening of others. In this way even mature networks are dynamic.

Networks increase the capability of individual firms. In this sense networks are knowledge. Faced with a problem firms tend to look within the network for answers about technology, products and useful contacts. They identify people who will cooperate and have specific capabilities. Consultants, contractors and specialists build up know-how about where to find key technologies, how to cooperate to develop new products, which research institutes are worth funding, who are strong competitors and what their current developments are. These strengths mean that innovation is facilitated by networks.

Partnering firms do not rely on traditional financial measures to guide their strategy. This is regarded as being as silly as driving a car by looking in the rear view mirror. Instead they measure customer satisfaction, operational efficiency and the involvement and commitment of their staff. An important consequence is that they pay dividends that take account of the needs of the long-term business and all the stakeholders. Dividends are typically a lower percentage of profits than is normal in firms wedded to the central importance of shareholder's interests. The actual amount of dividends is often higher in the long run because partnering helps generate higher profits. But partnering firms do not pay a huge proportion of profits in dividends because the money is needed to invest in building long-term strengths and it makes commercial sense to rely on internally generated funds. When they need to look externally for financial support they rely more on bank loans than on stock market finance. In doing this they tend to look for financial institutions that support local businesses long term.

Partnering consultants, contractors and specialists concentrate on their fundamental task. In construction this is to produce the best buildings and infrastructure. They integrate the interests of all the stakeholders to grow sustainably and make profits over the medium and long term.

5.7 Individual Training

Individuals may need training in communication, cooperative decision-making and technical issues to ensure that work teams are competent.

It is not easy for people who have learnt how to survive in the traditional industry to suddenly change. They often need training in cooperative teamwork. This can begin at partnering workshops and induction courses but needs to be reinforced by training.

The first step is to raise awareness of the benefits of training among senior managers. Their support should lead to a greater understanding and acceptance of the initiative throughout the firm. It often helps for the internal partnering team to appoint a senior manager experienced in and committed to using partnering to develop a training scheme.

The first step is to identify the need. This should include spending time with key people involved day to day in partnering relationships to identify areas of skills or knowledge that need to be fine-tuned or changed. At the same time the senior manager should market the benefits of partnering training by explaining how it will make work more successful and easier.

The next step is to discuss the apparent need with training organizations. Partnering facilitators can usually suggest suitable organizations. Training colleges and other academic institutions involved in construction management teaching and research should also be able to suggest suitable organizations. The aim is to establish what can be provided and the costs.

In devising the partnering training scheme, it needs to be understood that there are limits to training. Perhaps the most important limitation arises because highly developed skills are difficult to understand and copy because at least some of the knowledge involved is tacit; that is, embodied in the heads and hands of people, in teams, organizational structure, processes and culture. We know a certain practice works but we cannot explain and prescribe how. It has to be learnt from practice. There are even tougher barriers when an innovation is systematic so that a number of activities need to be changed in a coordinated manner.

For all these reasons training linked to work activities is usually most effective. It should be reinforced at induction courses and workshops. The training scheme should provide positive incentives for people to ask for training and reward them when they complete a course.

It is important that the scheme takes account of the general attitude towards training in the firm and fits into normal work patterns. In most consultants, contractors and specialist practices the internal partnering team will need to produce a business case that sets out the costs and benefits. Once the scheme is agreed and funded, the next step is to spread awareness of the partnering training scheme throughout the firm.

There is some merit in waiting until people are involved in a partnering relationship before they are given training. This is because many people find it easier to learn new concepts and attitudes when they can see the direct relevance to their work on a specific project. Nevertheless it helps if key people have appropriate training right from the start. They should be encouraged to talk about the benefits of the training and to take every opportunity to demonstrate what they have learnt. This tests the training scheme, gives others confidence in the benefits and encourages key people to look for training needs.

5.8 Developing Work Teams

Work teams must be empowered to make decisions about their work and its relationships with other work. This means work teams are provided with the information needed to fully understand the situations they face. Then when a work team agrees a course of action, it must have the authority to carry it out. Achieving all this may require work teams to have training in the use of quality, safety, time and cost control systems so that they can play a full role in project teams.

Work teams need to be equipped to play a full role in partnering. They do the work that delivers value for clients and earns profits for consultants, contractors and specialists. Organizations should do everything possible to enable them to work efficiently. In the vast majority of situations this means work teams being fully involved in making all the decisions that directly influence their own work. They need to work in cooperation with other work teams that influence or are influenced by their work. They need to work in cooperation with project core teams. They need to work in cooperation with the other teams inside their own firm.

These wide responsibilities mean that multi-skilled teams are crucial for partnering to deliver its full benefits. This means teams, in addition to their technical specialisms, need to be competent in a diverse range of modern skills. The specific skills required depend on project circumstances. They may include some or all of those listed in the box below.

Teams can be equipped with multi-skills by training or including experts in teams. The arrangements needed for training described in Section 5.7 should cover all the skills listed in the box below. In the medium to long term, training usually provides the best approach to establishing work teams with multi-skills. In the short term various ways of bringing experts into work teams are used in successful partnering. The simplest approach is for people with essential skills to be added to work teams. This can lead to unexpected benefits. Teams that include their own experts act more confidently in setting their own targets and finding ways to achieve them. As a result this apparently expensive approach may turn out to be the most efficient.

However, there are other situations where the benefits do not compensate for the direct costs. Some organizations deal with this by creating a pool of experts that are seconded to teams as and when they are needed. This is different from providing support services because the experts join the work teams for as long as they are

Skills in Multi-Skilled Teams

- Quality control
- Health and safety management
- Time planning and control
- Cost planning and control
- Risk and value management
- Managing change
- Procurement
- Legal skills
- Accounting
- Auditing
- Supply chain management
- Stakeholder management
- Relationship management
- Communicating
- Decision-making
- Problem resolution
- Benchmarking
- Feedback management

providing net benefits. This approach can help devise and disseminate good ideas as the experts exchange ideas that have emerged on individual projects.

Another approach that can work is for work teams to hire consultants as and when they need specific expertise. They play a similar role to an internal pool of experts. They bring ideas from a wider range of situations but may be less well focused on the firm's specific work.

The internal partnering team also needs to identify and develop the communication channels that will support work teams in using partnering. These need to ensure that work teams have absolutely up-to-date information about their work and where, when and how it fits into the project. In addition to using the communication channels to support their direct work, work teams should be encouraged to use the channels to exchange viewpoints, expectations, problems and new ideas. This all helps to give everyone a voice and a sense of ownership in the direction of the partnering arrangements.

5.9 Stability and Flexibility

Consultants, contractors and specialists need to combine stability with flexibility. Stability provides a basis for delivering efficiency. Flexibility enables the firm to support staff working in cooperation with people from other firms in a variety of partnering teams.

Construction projects require consultants, contractors and specialists to be flexible. Most projects have distinct characteristics that prevent a straightforward use of standardized solutions. So firms either develop a range of answers or concentrate on becoming skilled at dealing with one-off situations.

Partnering aims to help consultants, contractors and specialists improve their performance and so encourages the use of standards wherever they provide satisfactory answers because this leads to high levels of efficiency. Even when a firm can use standard technical answers, many project teams bring together at least some work teams that have not worked together before.

Many construction projects need original designs that give rise to many new relationships between work teams. Consultants, contractors and specialists specializing in such projects have to be flexible to support their senior managers and work teams working with what may be very different project teams. Partnering encourages the development of consistent processes and systems that support creativity and innovation.

This all means that consultants, contractors and specialists using partnering need to combine stability aimed at efficiency with sufficient flexibility to provide effective members of all the kinds of teams required by partnering.

Stability and flexibility require the directors to work with the internal partnering team to provide committed leadership. An absolutely key decision is which services are provided by specialist departments and which are part of work teams. It often makes sense for marketing, finance and human resources to be dealt with centrally. The rest of the organization should be a self-organizing network of teams. This is similar to the organization structure adopted by the most effective strategic arrangements described in Chapter 6. Using this form of organization internally helps work teams to fit easily into project teams that form part of a strategic arrangement.

Cooperation should be encouraged throughout the organization. Work teams, specialists, managers, suppliers and customers should be encouraged to suggest ways of improving performance. Anyone may suggest ways of improving quality, safety, time or productivity. Anyone may have ideas for technological or product developments or providing a better service to customers.

Self-organizing networks demand a range of personal skills and knowledge found only in top-calibre professionals. They spend little time on paperwork and concentrate on delivering value and developing key relationships. Partnering organizations aim to develop their staff at all levels so they have these characteristics and ensure they are suitably rewarded. This essentially is what the internal partnering team is there to achieve. Chapter 7, checklist 23 describes a set of principles that will help them concentrate on their essential responsibilities.

Partnering in Area 21

Case Study Reference: M4i 126

The Highways Agency, consultants Mouchel and contractors Accord Jarvis formed a dedicated team drawn from traditionally opposed camps to rewrite the rules on how to maintain and operate a highway network. As Mick Priest of Mouchel said: 'It's barely five years since consulting engineers like us got actively involved as highway managing agents.' He has arrived at an important conclusion: 'Anyone with delusions of being 'The Engineer' will not succeed in this business. I think we are witnessing the first steps in the ultimate merger of engineer and contractor functions into one company.'

'We were initially sceptical about the results we might achieve in partnering,' admitted Nick Atkinson of the Highways Agency. 'But the team agreed a charter in which they promise to work together for the benefit of road users. I truly believe that we are making near-optimum use of our resources to spend public funds as wisely as possible.'

Partnering in Highways Agency Area 21 arose after the Latham Report challenged the deep-rooted adversarial nature of the UK construction industry. Accord Jarvis's Barrie Groves emphasized the importance of leadership in responding to Latham's vision: 'Our first partnering workshop included two chief executives and other directors from the three organizations. They were enthusiastic and we felt convinced we were on the right road from day one.'

The first and most important action to address in establishing partnering was leadership. Senior managers established a set of principles that required attitude changes to cascade quickly throughout the whole team. An important part of achieving the essential changes was a series of director-led workshops for key people. These established the basis for workshops that brought the whole team together.

The partnering arrangement has provided unexpected benefits. Accord Jarvis and Mouchel saved the Highways Agency in excess of £500,000 by converting garages into modern joint offices that have played a key role in developing the partnering team. The facility also provides a long-term asset for the Highways Agency.

The main benefits of partnering in Highways Agency Area 21 included:

- Budget certainty – the team has objective performance measures that show it delivered the client's requirements within $2\frac{1}{2}$% of budget compared with cost over-runs of up to 15% that were common before partnering was adopted.
- Reduced administration – partnering with one contractor has radically cut administration costs for the client and consultants compared with traditional approaches that at times meant dealing with 14 agents and contractors.

Partners in crime

Case Study Reference: M4i 94

In the first major test of its partnering arrangement with Kent Police, Kent Property Services, the consulting arm of Kent County Council, has successfully produced the £5.7m Tonbridge Police Station within budget and 12 weeks ahead of the original 74-week programme.

Kent Property Services head, Geoff Rutt, explained their first actions: 'We engaged contractors very early and worked with Kent Police to develop the partnering charter. We then cascaded the same approach to every new organization and individual as they joined the project team and it works!'

According to Kent Police Force Estate Surveyor Bill Wallis: 'The contractor, Wates, pulled out all the stops to eradicate the old blame culture on this job and the results speak for themselves.' Kent Property Services head, Geoff Rutt, agreed and was adamant: 'Without their drive the project would not have been such a success.'

Wates site manager, Howard Sussons, was certain that involving suppliers and subcontractors as early as possible and listening to their advice made all the difference: 'That way you get people owning the end product and defects are greatly reduced.'

The principal partners nominated five champions to promote communication and leadership throughout the project team. The contractor partners assumed responsibility for developing partnering attitudes in their supply chains.

The benefits of bringing partnering into the firms were as follows:

- Costs were controlled more effectively than with any other approach. Wates' staff have expressed amazement at the many innovations suggested by suppliers and subcontractors simply because they understand the overall objectives.
- Time was controlled more effectively than with any other approach. 12 weeks were saved in construction and the police station was operational just 14 days after handover because the client's datacomms contractor was brought into the construction team.
- The client had the benefit of a predictable and early handover.
- There were few defects at handover, which is reflected in the project winning the 'Built in Quality' award from Tonbridge and Malling Borough Building Control.
- Safety was improved and there were no lost time accidents compared to the three normally expected on a project of this type.

Kent Police Force Estate Surveyor, Bill Wallis, summed up the team's views: 'Partnering is getting a quality job quicker, safer, cheaper and right first time. Construction is an enjoyable experience when everyone works together to achieve exceptional results for our clients.'

Partnering in a multi-project programme

Case Study Reference: M4i 068

NatWest Group Property developed a tailored partnering approach with its seven project teams responsible for creating a number of national centres to provide specialist banking services. The work formed a three-year building programme, comprising 60 projects valued at around £80m. The whole programme was completed on time and within budget.

NatWest Group Property began by assessing its existing portfolio of properties and formulating a £80m 'rationalization' programme. Through early internal workshops, the client was able to 'partner' and draw on the skills of the bank's own research and development and central services departments to determine:

- The partnering structure for the programme.
- The optimum number of teams and team members and configurations.
- The procurement method and major supplier agreements.
- The methods of performance monitoring.

Much time and effort were invested in creating teams. Seven teams were formed in which the key members stayed constant. The main aims were to meet tough time requirements and minimize risk to the bank's operations. With this in mind only medium-size companies that NatWest Group Property had worked with successfully and trusted were invited to tender. In forming teams, NatWest Group Property chose groups of firms who had worked well together on previous projects. They also selected specific individuals from within the short-listed companies as members of their teams.

NatWest Group Property appointed a partnering consultancy to help build on its own partnering culture and extend it to the seven project teams. The three main partnering objectives were shared benefits, measured continuous improvement and speedy non-adversarial problem resolution.

The key benefits of strategic partnering on the programme of 60 projects were:

- All projects delivered on time and within budget.
- No contractual disputes, despite significant logistics problems and time pressures.
- Fast resolution of problems peer-to-peer, without management intervention.
- Minimal risk to the bank's operations.
- Improved team-working and an atmosphere of trust throughout the teams.

Teamworking – the last frontier for measurement

Case Study Reference: M4i 120

Team effectiveness is one of the last measurement frontiers, according to Pearce Retail's head of human relations, Roger Leveson. He guided the use of a relatively new tool called the Team Climate Inventory to measure team performance on a demanding £1.5m project to redevelop a Safeway store in Chelsea. Leveson explains: 'The Team Climate Inventory measures the shared perception of how people feel about decisions, communications and work practices. On the Safeway Chelsea project the new measurement tool identified a number of areas where there seem to be areas where we could make improvements.'

The Team Climate Inventory uses a questionnaire that is completed by team members as frequently as they deem necessary. It asks about the atmosphere in the team, how people tend to work together, how frequently they interact, their objectives, and how much practical support is given towards the implementation of new and improved ways of doing things.

Pearce Retail's key account manager, Mark Giltsoff, explained that they found the Team Climate Inventory being used in the oil industry and by NHS management teams. He sees it as an important weapon in building effective teams: 'Use of tools like Team Climate Inventory will bring us into line with other sectors where measurement of soft issues is routine. The industry needs more hard and soft measurements if we are to raise our game, and that means gathering accurate data, and acting on it.'

Benefits of measuring performance with Team Climate Inventory on the Safeway Chelsea project included:

- Greater maturity – people had a mature understanding of behavioural change and attitude development.
- Team vision – the team had a clear sense of purpose and understood its collective strengths and areas for improvement.
- Structured feedback – team members had structured feedback on their performance and teamwork based on aggregated self-assessments.
- Innovation was encouraged – as an example, the Safeway Chelsea team devised a radical solution to overcome the problems of working on a congested inner city site.

Developing strategic collaborative working

6.1 Introduction

Project partnering delivers increasingly large benefits when it is developed long term. Strategic partnering is cooperative actions by a group of clients, consultants, contractors and specialists aiming to improve their joint performance over a series of projects. Strategic collaborative working is actions by a group of construction firms cooperating to develop a long-term business.

This chapter describes the emergence of strategic partnering and its further development into strategic collaborative working. The various strategic approaches are sometimes described as a strategic alliance or an inter-firm alliance. The terms strategic partnering and strategic collaborative working are taken to include these other terms.

Strategic partnering takes consultants, contractors and specialists beyond their traditional concentration on individual projects. Its purpose is to enable them to carry out projects effectively by acting and thinking long term. It provides more benefits than simple project partnering and enables consultants, contractors and specialists to deliver greater value to clients and earn larger and more secure profits.

It takes time to establish strategic partnering starting from traditional practice. The very significant benefits quoted in Chapter 1 took at least five years and many projects to become established in the normal practice of the clients, consultants, contractors and specialists involved.

Some groups of consultants, contractors and specialists have gone further and continue to find even more performance improvements. There is no obvious limit to the search for ever-greater benefits. However, it may take ten years and many projects to move away from traditional approaches through project management, project partnering, strategic partnering to strategic collaborative working and achieve the large benefits quoted in Chapter 1.

Construction's traditional approach which requires individual clients to assemble a project team to produce a new facility and then make their own facilities management arrangements is inefficient and outdated. Yet moving away from project-based methods provides severe challenges for many consultants, contractors and specialists. Most are small and many are run by individuals who value their independence, are naturally competitive and believe it is important to guard their knowledge and contacts. They tend to be suspicious of others and only cooperate defensively to fight off common threats.

Despite these conservative attitudes, the scale of change in demand, technology and business methods is forcing all consultants, contractors and specialists to rethink how they work. Some use informal local networks to think together about the best ways of tackling projects, solving problems and exploiting future opportunities. These tend to provide short-term answers but also identify longer-term problems. In searching for answers they often bring in external experts to help define and respond to longer-term needs. The external organizations typically include education, training and research organizations, various organs of local government, professional bodies and trade associations. As confidence grows some members become more creative in finding solutions to problems and grasping opportunities. Some of the informal links strengthen and develop into strategic partnering and on into strategic collaborative working.

Most strategic arrangements emerge as a natural development of successful project partnering arrangements. The consultants, contractors and specialists involved begin to feel inhibited by the limitations of project partnering. They value the benefits it provides and want to build on them. So they set up a strategic partnering arrangement that usually includes the client. However, some clients are not able to provide a sufficient flow of work to justify investing in strategic collaborative working. They may allow the consultants, contractors and specialists involved to extend the project partnering arrangements to other clients particularly if they are not direct competitors. In other situations consultants or contractors decide there are more benefits in breaking away from the original arrangement and using strategic collaborative working to provide a better service to a range of clients.

Another common trigger for consultants, contractors and specialists to adopt a strategic approach is pressure from a major client to work in a new region or overseas. This forces consultants, contractors and specialists to cooperate with firms that can provide the local knowledge needed in dealing with local regulations, local officials and local work practices. In return they introduce well-developed designs and processes into the local construction market. Such marriages can have considerable potential for building new businesses and the consultants, contractors and specialists involved may decide that strategic collaborative working provides the best way of seizing the opportunity.

Whatever the initial reasons for adopting strategic collaborative working, it is leading to the emergence of two distinct and highly efficient construction industries. The first is based on groups of consultants, contractors and specialists that together are very competent at tackling difficult projects. These tend to be large projects, which require individual designs, a creative use of new technologies and original construction methods. Many of them are further complicated by having to deal with challenging environmental conditions. These groups of consultants, contractors and specialists become skilled at dealing with the multitude of separate organizations involved in these major projects. The groups learn how to organize all the diverse knowledge needed to design and deliver innovative answers. They use strategic collaborative working to enable them to set up cooperative project teams that enable a client, all the stakeholders and a broad range of construction skills and knowledge to work together creatively and efficiently. They carry out these challenging projects with a confidence and certainty that is remarkable. The UK's Channel Tunnel Rail Link and Heathrow's Terminal 5 are important examples of the progress already achieved by groups of clients, consultants, contractors and specialists using strategic collaborative working.

The second distinct form of organization based on strategic collaborative working is groups of consultants, contractors and specialists that produce ranges of buildings or infrastructure facilities backed up by sophisticated client support services and marketed under brand names. The support services typically include finding land, providing finance, finding clever ways of helping clients understand design options and facilities management services. These groups of consultants, contractors and specialists reliably deliver good value products quickly and on time. These firms are building distinct sectors of the construction industry that have most of the characteristics of other consumer product industries.

These two distinct forms of strategic collaborative working are crucially important for UK construction. They have the potential to transform the industry and its reputation.

6.2 Strategic Approaches to Partnering

Strategic approaches to partnering involve a set of actions groups of firms take to improve their joint performance long term. It develops step by step as the benefits to the firms involved steadily increase. Strategic partnering and strategic collaborative working are the most distinct stages and are described in this chapter.

Strategic partnering exists when two or more organizations develop a close, long-term relationship based on working together to enable them all to secure the greatest benefits. The organizations accept that cooperative teamwork is more effective and efficient than competition. It works because the parties have an interest in each other's success. It works because it is based on the most fundamental reason for people to cooperate. This is not as so many commentators suggest that they trust each other, it is because they expect to work together again in the future. It is entirely natural for people who expect to interact in the future to cooperate. When people expect not to interact again, they look after their own interests. It is safe to trust people to behave in that fundamentally human way.

Strategic Partnering Defined

Strategic partnering is a set of actions taken by a group of clients, consultants, contractors and specialists to help them cooperate in improving their joint performance over a series of projects. The actions aim to agree an overall strategy, ensure the right firms are included, financial arrangements support partnering, firms' cultures, processes and systems are integrated, the most effective project processes are used, measured performance continuously improves and the whole arrangement is guided by feedback.

Strategic partnering develops over repeated interactions between firms as the people they employ learn to cooperate. It usually develops as an extension of project partnering. The actions taken by the people involved are guided by an agreed strategy and use feedback to ensure they continually improve their performance. The set of actions found in best practice is illustrated in Figure 6.1.

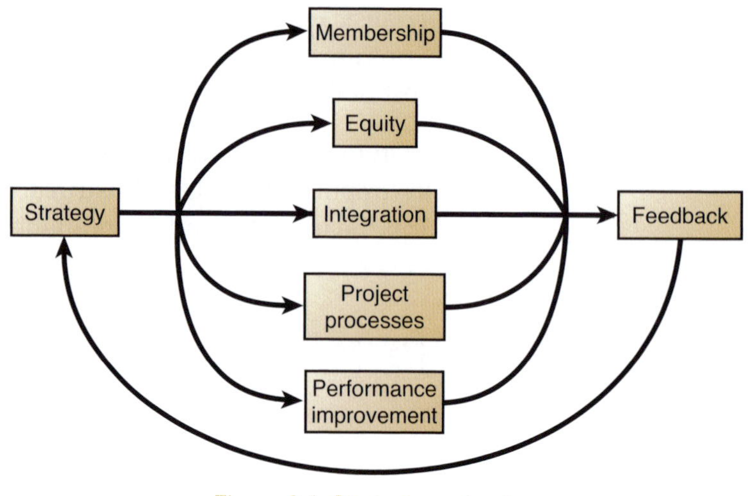

Figure 6.1 Strategic partnering

Further developments come when consultants, contractors and specialists already using strategic partnering decide to build a business on the basis of their collective strengths. They undertake market research to discover what clients in the particular market sector want, they research the relevant development processes and produce products and services to exploit the opportunities identified by these investigations.

The most highly developed forms of strategic collaborative working establish an integrated construction cycle. The broad elements of this most advanced approach are illustrated in Figure 6.2.

All strategic arrangements have a life of their own. They are guided not by detailed planning and control but by debate, tackling problems, seizing opportunities, trial and error, building on success and making changes over time in response to whatever seem to be the most important issues.

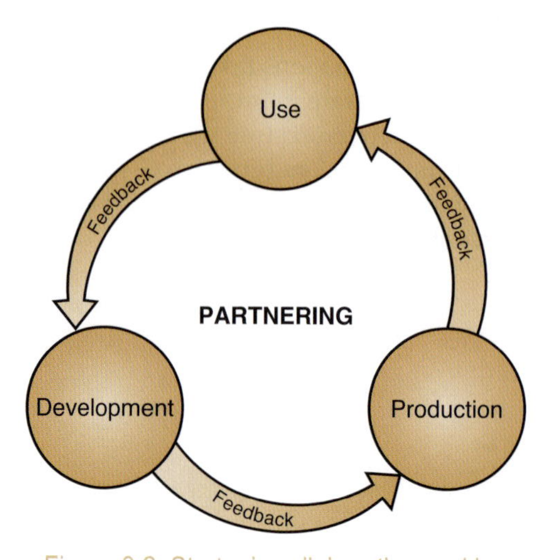

Figure 6.2 Strategic collaborative working

Strategic Collaborative Working Defined

Strategic collaborative partnering is a set of actions by a group of consultants, contractors and specialists to help them cooperate in establishing and continuously developing a long-term business based on an integrated construction cycle that links clients' use of constructed facilities with their development and production.

This chapter describes actions clients, consultants, contractors and specialists take to give themselves a realistic chance of making a success of strategic partnering and developing it into strategic collaborative working. The chapter begins by describing the benefits that can be delivered, to explain why firms are prepared to make fundamental alterations to the way they work to create a strong strategic arrangement.

6.3 Benefits of Strategic Approaches

The benefits grow as partnering moves through distinct stages. Project partnering can reduce costs by 30% and times by 40% compared to traditional approaches. Strategic partnering, which means a group of firms partnering over a series of projects, can reduce costs by 40% and times by 50%. Strategic collaborative working, which means consultants, contractors and specialists developing and marketing brand-named ranges of constructed products and services, can reduce costs by 50% and times by 80%.

Strategic approaches bring together complementary enterprises so that expertise, knowledge and skills are pooled. Communication between the partnering organizations is improved and duplication is reduced. Consultants, contractors and specialists share information and technical resources with clients and suppliers, which reduces uncertainty. This provides a framework that allows project teams to be kept together so they have time to improve the way they work, the facilities they produce and the services they provide. Costs and times are reduced, quality is boosted and responsiveness to clients' needs is improved.

Strategic approaches help consultants, contractors and specialists monitor and adapt to changing circumstances by identifying new services and products and responding quickly to market opportunities. They makes greater resources available to tackle large and difficult projects and new types of projects. Innovative ideas are more likely to be developed and used in new situations. As a result of all these benefits, consultants, contractors and specialists using strategic collaborative working have the strength and flexibility needed to expand into new markets, including international markets.

Clients involved in strategic approaches benefit from improved value for money and greater certainty. Cost and time overruns are a thing of the past because problems are addressed early and dealt with before they escalate into confrontation or claims. Quality is improved because project teams concentrate on eliminating defects. Projects are finished faster because lead times and production down time are reduced and project teams find better ways of working. Costs are reduced by eliminating duplicated activities, reducing administration costs, improving cash flow as well as greater efficiency at all stages of the project process. Risks are identified and allocated appropriately. In addition clients increasingly get the benefit of world-class facilities as the consultants, contractors and specialists they partner with help them find innovative responses to market opportunities and changes. This can provide a basis for a great variety of business benefits including more efficient production, lower running costs and better service to customers.

The benefits appear quickly for clients that need a series of similar facilities where there are known opportunities to drive out waste and inefficiency. In these circumstances strategic partnering can deliver massive reductions in time and cost while the quality and performance of the facilities produced is steadily improved.

Large or innovative projects where user requirements are difficult to specify and construction conditions difficult to foresee provide a very different challenge. Resources have to be assembled from a wide base. The resulting project teams have to learn to work together at creative problem-solving. Historically such projects run late, overspend and sometimes produce disappointing buildings or infrastructure. Clients, consultants, contractors and specialists that tackle these difficult projects can benefit from strategic approaches by developing techniques from project to project that provide greater certainty. The techniques help consultants, contractors and specialists set and achieve realistic budgets and programmes. This encourages clients to allow them the time and resources needed to produce outstanding buildings and infrastructure.

The data given in Chapter 1 shows that the benefits of strategic approaches increase over time as partners continuously search for better ways of working and improve their products and services. There appear to be distinct stages in these developments, the most significant of which are strategic partnering (which can reduce costs by 40% and times by 50% compared to traditional approaches) and strategic collaborative working (which can reduce costs by 50% and times by 80% compared to traditional approaches).

6.4 The Decision to Use Strategic Partnering

The first step beyond simple project partnering is a group of firms deciding to partner over a series of projects. The firms may be assembled by a major client or simply be consultants, contractors and specialists who have worked together successfully. The partnering firms set up a strategic team to lead the joint organization guided by the seven pillars of partnering (shown in Figure 6.1).

Common reasons for deciding to use strategic partnering include the following:

- An experienced client wants to work with selected consultants, contractors and specialists to improve their performance.
- A consultant or contractor has worked with other firms on several projects and decides that a deeper relationship could deliver more value.
- Someone spots a market opportunity and decides that working with other organizations provides the best way of producing the products and services needed to exploit it.
- Someone decides that their products and services need to be improved and this can best be achieved using the increased strength and flexibility that comes from working with other organizations.

Having taken the decision to consider cooperating long term, a strategic team comprising representatives from each organization needs to be set up to lead the use of strategic partnering. The team decides the first actions and ensures that arrangements are in place for them to be successful. It then monitors progress and decides on new actions and so on for as long as the arrangement continues. The strategic team's ongoing role is to keep driving the joint organization to higher levels of efficiency.

The strategic team's discussions should not be treated as contract negotiations where the parties take legal advice on detailed terms and conditions and embody the outcomes in formal legal documents. What is needed is an ongoing series of discussions amongst partners to agree joint actions that benefit them all. The outcomes should be recorded as a non-binding framework of principles that is regularly reviewed and updated. This provides the governing document, which describes how they all agree to act together in cooperation. It should not be turned into nor treated as a formal legal contract.

The following sections provide a checklist of actions that strategic teams need to consider. The seven sets of actions, shown in Figure 6.1, are sometimes called the seven pillars of partnering. Each describes a related set of actions that helps ensure second-generation partnering is successful.

6.5 Strategy

Strategic partnering is guided by an explicit strategy. It describes the type of buildings or infrastructure and services that will be produced and marketed. The strategy needs to fit the partners' internal organizations and be flexible to cope with change.

It is important that a strategic partnering arrangement has an explicit strategy, which describes a clear purpose and how it is intended to be achieved. This first pillar should identify the benefits each organization can expect to get out of the arrangement and describe how they will recognize and measure success.

A good way to start developing the strategy is for each of the organizations to think in terms of building on their own strengths and using partners' strengths to extend and reinforce their own capabilities. The questions in the box below may help the clients, consultants, contractors and specialists involved prepare for the strategic team's initial discussions.

The strategic team should begin by discussing the expectations of all the organizations involved. A good way of starting is to check that there is some agreement about the buildings or infrastructure it is intended to produce. It often helps to discuss how the new facilities will be used and how they may be used in the future. These discussions should take account of total life-cycle costs and environmental impacts. They should consider the benefits of flexibility that allow facilities to be used in different ways. New ideas can be generated by considering developments in technology including the use of prefabrication and standardized components and their potential to improve quality, time and cost.

Once they have agreed the kind of buildings or infrastructure to be produced, the strategic team can consider the clients it intends to attract. They should particularly consider how clients' interests will be taken into account. It helps to focus on quality and certainty of delivery.

Questions to Help Prepare for the Strategic Team's Initial Discussions

- What is your organization good at?
- Can any of these strengths be developed?
- What new strengths can be developed?
- Which existing or new strengths provide a promising basis for a new venture?
- Does the new venture fit your existing strategy?
- Will there be problems getting your organization to agree to the new venture?
- Will there be problems developing your organization to suit the new venture?
- Will the new venture strengthen your position in an existing market?
- Will it secure new business?
- Will it protect you from adverse trading conditions or increased competition?
- Will it improve profitability?
- Which of these issues can the other strategic partners contribute towards?

This sets the framework for the strategic team to agree the marketing activities to be undertaken. It is sensible to consider which markets provide opportunities for developing existing products or services and which provide opportunities for new ventures. It is important to consider whether the strategy exposes partners to more intensive competition and if so how they should prepare.

The strategic team should consider the joint organization's market image. This is inevitably influenced by its attitude towards corporate social responsibility including the way human resources are treated. This means considering employment terms, training and development policies and pension provisions. The market image is also influenced by the environmental impacts of the joint organization's activities. This means considering attitudes to waste disposal, especially hazardous waste, the impact on vulnerable ecologies, local communities and economies. It may well be sensible to support local development programmes that help the community. Amongst other benefits this may bring goodwill for future projects. Helping provide social housing can be an effective way for consultants, contractors and specialists to build a good local reputation.

In all this the strategic team should consider how their ideas fit the partners' existing strategies and objectives. They should consider whether a strategy being considered will strengthen the organizations or make them vulnerable to external forces or threats. This strategic risk analysis provides information that is vital in persuading senior managers to provide time and resources for the strategy to be produced, discussed, improved and agreed.

As the strategy is developed it should be discussed widely inside all the organizations involved to build support at every level. It is sensible to bring suppliers and clients into the discussions to explain the emerging ideas, discuss any queries they have and to assess their interest and commitment. When the strategy impacts an existing business it is likely to be greeted sceptically and may well generate opposition. So selling it can be tough.

The formal decision usually depends on identifying clear savings and performance improvements. These may come from design and construction innovations, organizational changes including reducing the number of suppliers, simplifying project processes and outsourcing activities to stronger partners. Risk can be reduced by allocating and managing it appropriately. Firms may be able to increase the scale of their operations in ways that deliver greater efficiency and profitability.

Strategic partnering also has costs, the majority of which arise in the early stages. Costs may rise due to increased complexity if partners with different cultures and business practices are asked to cooperate. There may be a need for training in communications and new technology, particularly if the new approach faces some firms with a scale and intensity of competition different from what they are used to. Rapid growth, or new methods of working, can give rise to unusual costs. In the long term the benefits outweigh the costs but they must all be taken into account in developing a viable strategy.

The agreed strategy should set specific short-term goals and objectives. It should identify what needs to be done and how it will be done in detailed step-by-step processes. The plans should also take account of the long term and at an early stage the strategic team should establish more ambitious goals and objectives. In this way the strategy provides some quick 'wins' as well as promising much more in the future.

The strategy needs to be sufficiently flexible to cope with changes. Different economic or market conditions, new technologies, or more demanding clients may mean the strategy has to be changed. More fundamentally, as the joint organization achieves its early goals it is likely to become more ambitious and want to go further and faster. To provide for this the strategy is reviewed regularly by the strategic team to ensure that it remains relevant and still challenges the partners to find ever better ways of working.

6.6 Membership

Partners are chosen because they want to work together. Partners are chosen on the basis of their track record in providing the technological skills and knowledge and the business characteristics needed to pursue the strategy. Membership is reviewed regularly and necessary changes made openly, generously and sympathetically.

The membership pillar is concerned with the choice of partners. The guidance given in Chapter 2 applies to strategic partnering but some issues need particular care.

The process of selecting partners can be initiated by any of the partners although most arrangements are started by experienced clients. It should be kept in mind that partners choose each other because they want to work together.

The choice of partners should be based on a clear understanding of each other's expectations and goals. This develops throughout the process of choosing the members of a strategic partnering arrangement, which inevitably takes place in parallel with developing the strategy. This is because the strategy establishes the kind of organizations needed as members, and the partners' aims and ambitions determine the strategy they will buy into. This in turn determines the technological skills and knowledge, and business characteristics that partners must have. As well as design, manufacturing and construction, mature strategies usually require partners to provide operational and maintenance capabilities.

All decisions about membership should be based on carefully researched information about potential partners' track record. The list provided in Chapter 7, checklist 6 provides a good starting point in establishing the information needed.

The information used to select partners should identify whether the people who will be involved can manage relationships as well as deal with paperwork. They need cooperative attitudes. They should be more skilled at forging agreements than dealing with disputes. Getting the right people in place and well motivated may mean some of the organizations need to find new recruits, invest in training or adopt better salary and reward systems. They may need to plan distinctive career patterns for people skilled in partnering if their main business is based on traditional, adversarial contracts.

Ideally, partners' strengths complement rather than duplicate each other so the joint enterprise can build on existing strengths. It is important that the organizations' growth and other strategies match each other. Size and turnover are irrelevant if the organization is the right partner. Many of the best consultants, contractors and specialists are small. They can be very effective partners and will develop and grow given the opportunity of working on large projects.

As the partners gain experience of working together, the strategic team should regularly review the contribution of each organization and actions taken to encourage a continued commitment to improving joint performance. The reviews should be provided with objective information produced by formal feedback systems based on the guidance in Chapter 7, checklist 15.

Even when this is all done well, there may come times when partners have to be changed. The change may result from any one of many possible reasons. A key partner may decide to expand a successful UK-based business overseas. They may be taken over by a major company with different ideas about how the business should be run. They may want to make more effective use of information and communication technologies. They may want to replace site-based construction processes with manufactured modules so that completion times can be reduced.

When a partner decides a change is needed this is nearly always a major shock, leaving some individuals feeling badly let down. It is important that changes are handled in ways that avoid bad outcomes.

The first requirement is that everyone involved recognizes that strategic partnering is effective only as long as it is in the partners' best interests to cooperate. When that is no longer the case, changes will be made.

The second requirement is that partners are alive to changes in partners' strategies, clients' demands, competitors' initiatives and developments in relevant technologies. These should be responded to positively so the strategic partnering arrangement is strengthened. Then if changes are needed, they do not come as a surprise.

The third requirement is that when a change requires new skills and knowledge, the partners realize that those unable to respond will be replaced. This very tough decision should be based on a rigorous evaluation of partners and potential partners.

The fourth requirement is that the need for change is discussed openly and any partner required to leave treated generously. All their real concerns and problems should be dealt with sympathetically. It is in everyone's best interest for changes to be agreed as amicably as possible in a spirit of cooperation. Fair treatment sends important messages to the remaining partners that they too will be treated well if changed circumstances in the future mean they have to leave. It also helps maintain a good public reputation particularly with clients, suppliers and potential new partners.

6.7 Equity

Financial arrangements must be designed to encourage partnering. Clients should get better value than is available anywhere else and consultants, contractors and specialists should get higher than normal profits. Initial finance may have to come from internal savings. Long-term investments may attract financial support from government. Financial arrangements should be open-ended to encourage partners' commitment.

Financial arrangements agreed at the outset should encourage the organizations involved to invest in long-term development work aimed at improving their joint performance. The equity pillar aims to achieve this by ensuring that the financial rewards from a strategic partnering arrangement are better than firms could get operating outside it.

Benefits should be shared in a way that accords with the partners' expectations. This is unlikely to mean an equal division even if it were possible to determine what that means in the context of a series of construction projects. Profit levels for consultants, contractors and specialists vary widely. The benefits enjoyed by clients derive from the impact a new facility makes on their business. The aim therefore should be to ensure that the financial arrangements are regarded as fair by all the partners. This means that architects, for example, should expect to do as well as any other architects would do in the same situation. They should not compare themselves with specialist contractors or developers. They have different financial structures and different risks that justify different rewards.

It is unlikely that financial arrangements will be perceived as fair if the parties think in terms of normal buyer–seller relationships in which outcomes are determined by size or power. The parties need to discuss what they regard as a fair basis and what they expect to get from and contribute to the arrangement. It sometimes helps in the early stages of a strategic partnering arrangement to adopt simple financial arrangements. These might mean sharing the additional costs of using partnering and the resultant savings on a predetermined basis. However, the fixed price elements of these simple arrangements tend to cause problems and even disputes. This is why successful strategic partnering tends to move fairly quickly to more sophisticated financial arrangements.

Arrangements that work in the long term are usually based on agreeing prices for new facilities that represent better value than the client can get anywhere else in the market. Then, within the agreed price, the consultants, contractors and specialists involved are guaranteed all their direct costs plus an agreed profit and contribution to fixed overheads. This too is more generous than they could expect from any project-based arrangement. Over time successful strategic partnering makes it possible to improve the initial deal so that clients get ever-better value and consultants, contractors and specialists earn ever-higher profits. In many cases clients choose to use some of the benefit delivered by strategic partnering to improve the quality of their new facilities. Similarly consultants, contractors and specialists often invest in training and new systems and equipment.

In addition to internally generated funds, the long-term perspective required by strategic partnering can help consultants, contractors and specialists gain access to external finance. They can look to government R&D funding to support development work. Various schemes exist and they usually involve joint work between firms, universities and other research and development organizations. Also the long-term view inherent in strategic partnering has obvious links with private finance initiative and private–public partnership schemes as described in Section 6.14.

Whether finance is generated internally or externally, the financial arrangements should be based on open-book accounting that gives all the partners access to each other's accounts. It may help in the early stages of strategic partnering to draw up agreements to ensure confidences are kept. As clients, consultants, contractors and specialists work together, the formal contract will become redundant as people learn what they can expect from each other. It is important that everyone is confident about the accuracy of the costs, profits and performance improvements used to measure financial outcomes.

The financial arrangements should give all the partners confidence that the relationship is on a sound footing and will continue for the foreseeable future. There must be no feeling that one partner is being exploited. Problems, worries and concerns must be discussed frankly and constructively. They should be seen as opportunities to make improvements, not signals that the relationship is about to end.

This open way of dealing with financial issues is likely to raise concerns and problems inside some organizations. It challenges the Anglo-American model of business that sees ruthless competition as the best way of achieving the paramount objective of increasing shareholder value in the short term. Strategic partnering is closer to the Japanese and European model of business based on social inclusion by which companies have long-term responsibilities to shareholders, the workforce, suppliers and local communities. Both models have strengths but they are different and strategic partnering is more long term than short term and closer to a social inclusion than a shareholder value model of business. It is often the process of spelling out these implications in financial terms that provokes the fiercest opposition inside some partners' organizations.

Some of the opposition may be well-founded. Firms involved in strategic partnering may face increased financial constraints because they are operating with partners. Funding may become more difficult to obtain. Financial institutions may place restrictive conditions on loans. They may be unwilling to take account of potential savings from a commitment to pool resources. They may not accept that financial benefits will result from risks being shared appropriately. These cautious attitudes mean it is not uncommon for the initial stages of strategic partnering arrangements to be funded from cost savings. This can make it very difficult to produce a convincing business case and it is likely that potentially beneficial arrangements are killed off before they begin.

Short-term attitudes are reflected also in suggestions that partners should agree an exit strategy at the beginning of a strategic partnering arrangement. This is not a good idea because agreeing the terms and conditions that will apply introduces negative, adversarial attitudes into the early discussions. More importantly if the parties know how the arrangement will end, there will come times when they will be tempted to calculate the short-term advantages of ending it. It is the open-endedness of relationships that causes parties to cooperate and continue searching for mutual benefits.

6.8 Integration

Partners' organizations and processes are integrated. Information and communication technologies and modern forms of face-to-face meetings play key roles in this. The strategic team provides leadership in making changes and keeping people informed about what is happening and why.

Strategic partnering delivers improved performance by integrating activities traditionally kept separate. The integration pillar aims to blur the boundaries between the operations of partners as activities are integrated into efficient delivery systems. Design, planning, construction and completion are treated as one integrated system in a continuing drive to eliminate waste and inefficiency. Supply chains that feed into the design and construction processes are integrated so the number of links is reduced and the chains are made more efficient. Production and innovation are integrated by balancing necessary rules and procedures with creative freedom.

Integration has to have a context so people can see the reason for changes to systems and procedures. A major project or a series of projects can provide opportunities for partners to integrate their operations. Whether the resulting new entity is given formal legal status or remains a virtual organization depends on the wishes of influential people in the organizations involved.

Integration requires top management to ensure that strategic partnering is supported by all departments. This is easier to achieve when the partners are experienced in partnering. Even with that advantage, new strategic partnering arrangements raise challenges for people and departments not directly involved in the construction projects. It needs leadership from the top to help deal with problems encountered as all the joint business processes are reviewed in a search for ways of integrating them. This means eliminating duplicated activities, concentrating on partners' strengths and cutting out waste whatever form it takes. It also means using the cross-fertilization of ideas that comes from operating across organization boundaries to encourage innovation.

The resulting changes inevitably change people's roles and responsibilities. This is one reason why excellent communication is important. People need to know what is happening and why. Communication systems need to ensure that needs and expectations are continually addressed. They need to ensure that performance indicators are monitored and problems spotted early so they can be resolved quickly in a non-confrontational manner.

Strategic partnering makes all aspects of communication more important. It is common for information systems and face-to-face communications to be redesigning to foster cooperation.

Information and communication technology is speeding up all business processes. Project networks ensure that everyone is working on absolutely up-to-date information. Strategic partnering links project teams so people faced with problems can identify everyone in the joint organization with potentially relevant information. In this way strategic partnering is fundamental to construction, realizing the full benefits of information and communication technology.

In parallel to these technological developments leading practice makes better use of various kinds of face-to-face meetings. A real effort is made to ensure that formal meetings have clearly defined purposes that everyone attending knows in advance. People are expected to know the background and be properly briefed on the key decisions that will be discussed and agreed. Workshops and task forces play central roles in strategic partnering. Social events are used to build team spirit and give people a rounded picture of their colleagues so they can communicate more effectively. Team offices that bring people from the client organization and different consultants, contractors and specialists together are used for crucial stages of individual projects. They encourage open decision-making and provide opportunities for everyone to join in discussions. Chapter 7, checklist 21 provides more ideas about face-to-face meetings and electronic links.

6.9 Project Processes

Project processes are made efficient by standardizing on best practice and looking for improvements outside individual projects. The resulting standards provide for well-developed designs, technologies and methods or processes that encourage creativity and innovation.

It is fundamental to strategic partnering that project processes are efficient. All the pillars of partnering contribute to projects being carried out quickly and reliably to high standards.

The project processes pillar aims to provide project teams with standardized actions and technologies that represent current best practice. An important part of achieving this is to constantly search for further improvements in performance, test them and incorporate the best in the standards for use on future projects.

For projects that use well-established answers, standards should enable project teams to be assembled quickly and well-drilled work teams to carry out their work virtually automatically. Most ideas for improvement will be identified by the strategic team reviewing feedback from several projects. Ideas will be explored and if they appear likely to make a significant improvement to project performance, developed by a task force. Having been thoroughly researched and tested, they will be tried out on one project that will be closely monitored to ensure that the intended benefits really exist. When the strategic team is sure the new idea has value it is incorporated into the standards and work teams are given any training needed to ensure the change is applied correctly.

The results are that project teams are not distracted from efficient work by new designs or changes. They concentrate on working efficiently to produce high-quality buildings or infrastructure that provide what clients want and give their own firms high profits.

For projects that require an original design, project teams need a flexible suite of processes designed to encourage creative and innovative design. These support talented, individual designers. They bring specialist technical and construction expertise into the team as and when it is needed. They provide state-of-the-art computer-aided design and communication systems. They support project offices where teams from different organizations can work together. They allow clients to set realistic budgets and completion dates and give them the assurance that they will be achieved. They provide rigorous quality control systems at all stages of the project. They encourage wide discussion and questioning of design and construction decisions. They ensure the use of non-adversarial, cooperative approaches. They encourage the use of established answers where they exist. They provide support when ideas for improving the project processes are identified.

New ideas may be developed and applied during individual projects. Alternatively, ideas that cannot be used on a current project are reported to the strategic team. All the ideas for improving project processes are reviewed by the strategic team. Any that offer real benefits are incorporated into the best-practice procedures. In some cases they need to be further developed by a separate task force. This approach enables project teams to deliver new facilities based on original designs, the best of which delight clients and contribute positively to local communities and the environment. It also ensures that consultants, contractors and specialists make fair profits.

6.10 Performance Improvement

The strategic team ensures the joint organization makes measurable performance improvements. The whole point of partnering is to achieve continuous performance improvement. It helps to establish benchmarks that can be used to set targets for improvement and provide the basis for feedback.

The performance improvement pillar guides the strategic team in ensuring that the joint organization continuously makes progress towards strategic objectives. It helps if current performance is measured in ways that can be related to the partners' previous performance, industry norms or best practice. This provides benchmarks that should provide information about the overall performance of the strategic partnering arrangement, the achievements of individual projects and the progress made by partners in adopting partnering throughout their organizations.

Benchmarks should be easy to record and should measure real improvements that benefit the partners. They should show what is delivered to clients compared with what is available in the same market sector. Typically benchmarks deal with the quality of outputs, the speed and certainty of their delivery, the incidence of defects, how quickly they are dealt with and prices. Benchmarks should include measures of client satisfaction with the product and service. They should measure the experience of users and neighbours. It is essential that benchmarks make sense to the client.

Benchmarks should measure consultants', contractors' and specialists' performance compared with competitors in the same sector of the construction industry. They should measure completion on time and budget, quality, productivity, safety, construction costs and times, and profitability. They can also usefully measure the elimination of waste, the quality of communications, staff turnover, investments in systems, training and equipment, and the number and effectiveness of innovations.

Benchmarks are used to set targets for improvement for the strategic partnering organization, project teams and the partners' own organizations. These should challenge the various parts of the organization to continually develop and improve their processes and products.

A good way of establishing benchmarks is for each of the partners to consider the best way of measuring those elements of the agreed strategy that most concern them. The outcomes should be discussed by the strategic team to find the most effective and practical measures. Increasingly the key performance indicators promoted by Constructing Excellence in the Built Environment and listed in the box on the next page provide the best starting point because they provide well-established measurements and are supported by a significant body of performance data.

Other aspects of construction that could be measured include user satisfaction, contribution to the local community and environmental impacts. Once an initial set of benchmarks is agreed, it helps to run trial projects to check that the resulting measures provide useful information.

Constructing Excellence in the Built Environment's Key Performance Indicators

- Client's satisfaction with the product
- Client's satisfaction with the services
- Defects
- Costs
- Predictability of costs
- Time
- Predictability of time
- Safety
- Productivity
- Profitability.

It helps if teams measure their own performance and the results are plotted graphically in simple diagrams that relate to agreed targets. Results should be discussed as soon as they are available and teams encouraged to suggest ways of improving their performance. All suggestions should be taken seriously and any that require actions by senior managers should be dealt with quickly and decisions reported to the teams immediately.

It is crucial in establishing good benchmarks that the strategic team keep in mind that their purpose is to drive the partnering organization towards performance improvement. It is easy to fall into the trap of measuring for the sake of measuring and lose sight of the overall purpose, which is to improve performance.

With this in mind, the strategic team should regularly consider whether the benchmarks need to be changed to reflect developments in the strategic partnering arrangement. They should check that their objectives, targets and plans remain relevant and mutually beneficial and that the planned resources and timeframes are still realistic. The reviews should help them identify opportunities to extend their activities or to recognize that they have reached a critical stage when further improvements are uneconomical and they need to make a step change in the organization's activities. As the strategy evolves in this way, new benchmarks will be needed that provide the basis for a new series of performance improvements. Most likely they will deal with long-term targets that may include greater market share, new product development, greater added value and environmental concerns.

6.11 Feedback

Strategic partnering depends on feedback systems to provide information on the performance of the joint organization, individual partners and projects. The strategic team ensures that feedback is effective and looks for ways of improving it. They use feedback to review the strategy.

The final pillar deals with the actions needed to provide feedback for the strategic team. This enables the seven pillars to act as a controlled system as illustrated in Figure 6.1. The strategy devised by the first pillar gives overall direction to the actions resulting from the next five pillars and the whole is guided by information about achievements and performance provided by the feedback pillar.

The strategic team should establish feedback systems that tell them whether their strategic objectives are being met. The systems should provide measured information about progress and performance and highlight problems and opportunities. They should measure whether teams are using partnering effectively. They must be accurate and timely so the strategic team can decide if changes are necessary.

In addition to the strategic feedback, individual organizations should have a programme of internal reviews to assess progress and the continued relevance of the strategic partnering arrangements. It helps if all partners measure the benefits they are getting from the strategic partnering arrangements on a consistent basis. They should also identify concerns, problems and ideas for improvement.

Project processes should include systems for measuring project performance that guide project teams towards their objectives. A basis for effective project feedback systems is described in Chapter 7, checklist 15.

The strategic team should not rely only on formal feedback but should regularly visit the offices, factories and construction sites where the joint organizations' work is under way. They should ask questions of the people doing the work and be prepared to answer their questions about the strategic partnering arrangement. The aim of the visits is to provide first-hand knowledge to help them interpret information provided by feedback systems. Chapter 7, checklist 15 includes questions that strategic team members can ask during visits to workplaces.

The strategic team should regularly review feedback reports covering the performance of the joint organization, individual partners and projects. The feedback should provide objective measures of how close the joint organization is to meeting all its targets in terms of agreed benchmarks. This should help establish what is working well, and identify problems. Thus if feedback shows that the joint organization is failing to meet a target of zero defects, the strategic team needs information on defects and their causes. If feedback shows that the joint organization is failing to meet a target for fast construction, the strategic team needs information on deviations from planned progress. The strategic team should take decisions to solve problems quickly, monitor overall progress, fine-tune strategic objectives and encourage the partners to continue searching for ever better ways of working.

It is important for the strategic team to have feedback from their buildings and infrastructure in use. This should come from people responsible for running and maintaining the facilities, users (including where appropriate customers), and the owners.

The specific outcomes of the strategic team's reviews should include targets for project teams and task forces set up to tackle specific problems or exploit opportunities. The guiding principle is that the strategic team's feedback-driven decisions should influence future actions in ways that deliver performance improvements.

6.12 Strategic Partnering Organizations

Different strategic teams make different decisions under the seven pillars (shown in Figure 6.1) which give rise to distinct organizations. They usually include a strategic team, project teams and task forces and may include interface teams and internal partnering teams. The overall organization should form a self-organizing network in which useful links become strong and the whole is guided by feedback.

The strategic team needs to establish an organization to put their decisions into effect. In most cases this is a virtual organization rather than a distinct legal entity. Decisions guided by the seven pillars of partnering result in different strategic teams devising very different strategies, objectives and ways of working. These various outcomes mean that strategic partnering organizations come in a variety of forms. Chapter 7, checklist 33 provides a glossary that describes the various kinds of teams used in modern construction.

In the most successful arrangements, the clients, consultants and contractors involved expect to work together long term because they believe they will get more benefits inside the strategic partnering arrangement than they would outside. In practice most strategic partnering arrangements do not last indefinitely. Circumstances change, new competitors emerge, clients' needs change or one of the partners gets new managers with different ambitions. It is not inevitable that any of these circumstances will arise, nor is it certain that if they do this will signal the end of the strategic partnering arrangement. The future is uncertain and it is sensible to plan for strategic partnering to be successful. This means that the organization set up to put the arrangement into effect looks and feels permanent. This is no different from the situation facing individual organizations, all of which face an uncertain future but sensibly plan as if they will survive forever.

Successful strategic partnering organizations are likely to include some or all of the elements shown in Figure 6.3 and described in the following subsections. Strategic teams should use the diagram as a

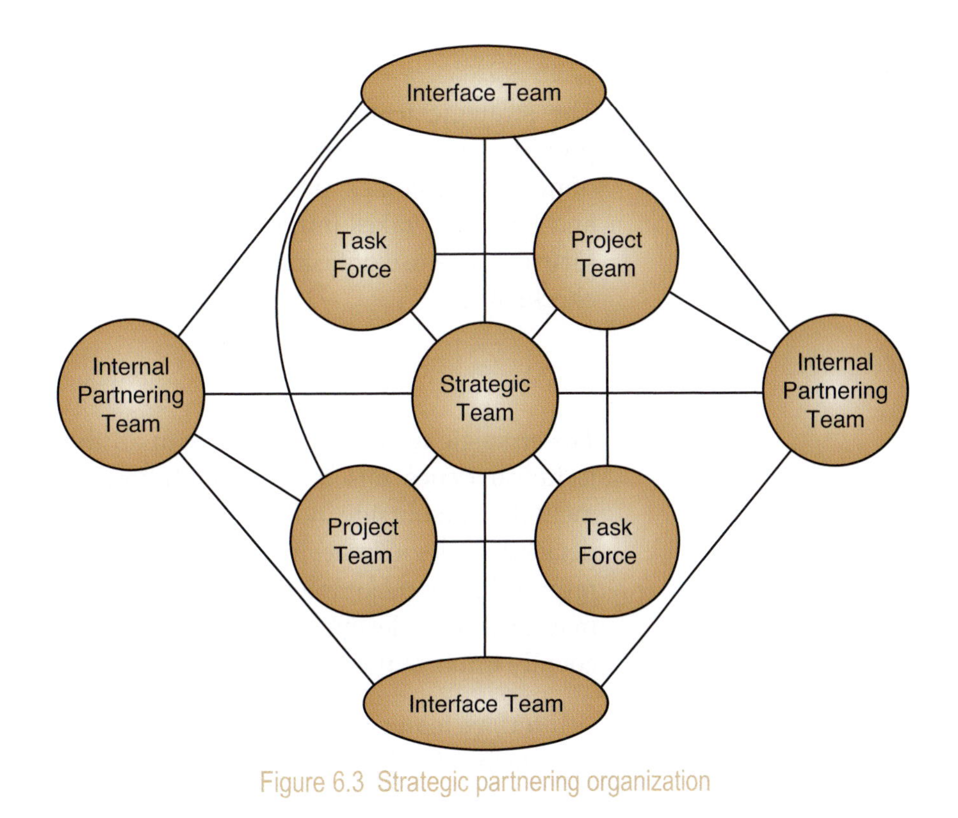

Figure 6.3 Strategic partnering organization

checklist to ensure that all the essential interests are incorporated in the overall partnering organization.

Strategic team

As described in Section 6.4 the first element of a strategic partnering organization to be set up is the strategic team. It comprises senior managers from all the clients, consultants, contractors and specialists involved in the strategic partnering arrangement. It provides overall leadership and makes strategic decisions, which are decisions that go beyond individual projects.

It is important for the members of the strategic team to have the authority to make decisions without having to refer back for approval. The members of the strategic team need to be able to rely on each other to act on the basis of joint decisions in a manner that supports agreed objectives and long-term goals. They need to be able to discuss issues as they arise in an open and positive way and make decisions to prevent minor differences becoming major crises. They need to be able to rely on each other to be open about situations where individual interests conflict with joint decisions that are being considered. Any such situations should be discussed until an agreed answer is found that all the partners support and believe is fair.

Project teams

The strategic team sets targets for project teams so that performance steadily improves and new ideas are introduced and used effectively. Individual project teams use project partnering as described in the earlier chapters of this code of practice to achieve the strategic targets.

Task forces

The strategic team sets up task forces to solve specific problems, provides the basis for major process improvements, explores the use of new technologies, designs new kinds of buildings or infrastructure and generally takes initiatives that drive the partnering arrangement forward. The use of task forces in partnering is described in Chapter 7, checklist 22.

Interface teams

The strategic team cannot deal with all the specific, detailed and often technical issues raised by an ambitious strategic partnering arrangement that fall outside individual projects. One way of dealing with these detailed strategic issues is to establish interface teams. These are teams responsible for major interfaces between the parties. Thus interface teams may deal with design, technology, quality, time, cost, safety and other equally important issues that affect the partnering arrangement.

Interface teams comprise the managers in each partner organization responsible for their contribution to a specific interface. The teams should meet regularly to deal with problems, identify and introduce improvements to products and services, and devise innovations that provide the basis for strategic changes.

Internal partnering team

Each partner firm should establish an internal team comprising all the interface managers plus representatives of any other key internal interests. The relationship between interface teams and internal teams is shown in Figure 6.4 to show how partnering gives individuals both internal and external roles.

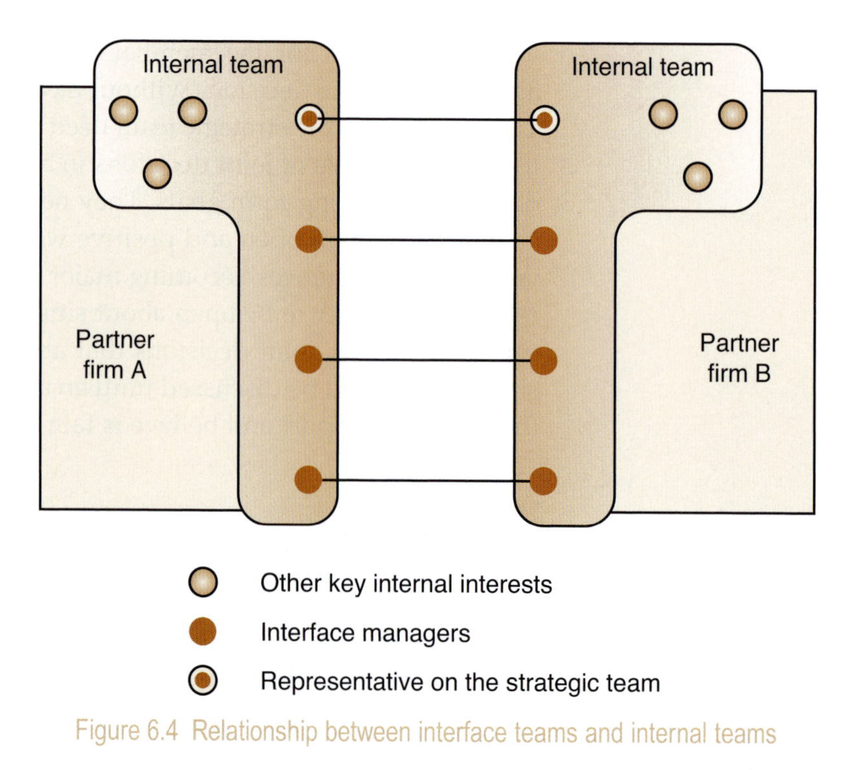

Figure 6.4 Relationship between interface teams and internal teams

The internal team should be a dynamic force ensuring the organization is playing a full role in the strategic partnering arrangement. It should meet regularly to check that there is genuine commitment to the partnering arrangement from top management. It should ensure that there is consistent effort towards sustaining the relationship and in particular that the organization keeps all its promises. It should ensure that there are sufficient resources in terms of the quality and number of people and equipment to do everything required to produce agreed outputs.

Overall organization

The various teams that are set up should not be arranged in a management hierarchy. They should all be empowered to communicate with whichever other teams they find useful. Information and communication technology should be used in ways that encourage the free and open flow of information. Joint meetings, including formal workshops and exciting social events, should be used to build understanding and commitment to the strategic partnering arrangement.

The right model for the overall organization is a self-organizing network. Links that teams find useful become strong. Feedback guides decisions, which should be made close to the action where all the relevant information is available and properly understood.

People are rewarded for committed work. Mistakes are seen as opportunities to find better ways of working. Time is never wasted trying to allocate blame. There is a relentless drive to succeed by competent people excited by the chance to do their best work. They understand that they all succeed by ensuring that projects are delivered in ways that delight clients, consultants, contractors and specialists.

It is important that these highly efficient organizations give explicit attention to providing for clients to be fully involved. This is important in all projects but is absolutely vital where the organization produces original designs to meet specific client needs.

6.13 Construction as a Creative Industry

Strategic partnering is leading to the emergence of groups of consultants, contractors and specialists that combine creative design with the skills needed to produce the highly individual buildings and infrastructure in ways that meet the needs of individual clients. This provides one form of a distinctly new way of working called strategic collaborative working.

Developments in strategic partnering are leading to the emergence of groups of consultants, contractors and specialists that are highly competent at producing a broad category of individually designed buildings or infrastructure. They use highly developed forms of strategic partnering taking account of all seven pillars of partnering in agreeing how they will work together. Some of these groups decide to use their highly developed competences to establish a long-term business.

The partners develop a deep understanding of the business aims of clients in distinct sectors of the construction market. This involves working closely with potential clients either in the public or private sector to develop ever more useful descriptions and measures of the business aims served by new construction. For highway authorities this could involve developing measures of safety, reliability, average speeds, warnings of hold ups and other ways of defining the performance delivered to road users. For a major sports stadium this could involve measures of the performance of teams using the stadium; the level of enjoyment experienced by spectators; regional, national and international impact in the media; and other factors that determine success from the point of view of the client.

When the partners win a project, they provide a team of creative designers, technologists, managers and specialists to work in partnership with the client. They actively search for the best possible design that delivers the most value consistent with the client's budget and timescale. This often involves developing new designs and innovative technologies. It may mean working out how to complete a project faster so the client can seize a new business opportunity.

The partners develop processes and techniques that encourage creative work. They agree realistic budgets and completion dates and know how to meet them. They bring all the stakeholders into the design process and look for good ideas wherever they happen to arise. They have well-developed supply chains skilled in helping to develop new answers and turn them into the elements and systems of buildings and infrastructure. Quality, health and safety are integral to all the processes. The partners take pride in delivering complex new facilities fully complete on time. These most effective consultants, contractors and specialists provide an important form of third-generation partnering.

6.14 Public–Private Partnerships

Public–private partnerships, which include private finance initiative projects, should use strategic collaborative working.

Public–private partnerships, which include private finance initiative projects, are one of the three approaches recommended for central government construction projects. The others are design build and prime contracting, which are described in Chapter 7, checklist 3.

Public–private partnerships establish long-term relationships between public authorities, consultants, contractors and specialists. An organization is employed to produce a new constructed facility and then take responsibility for delivering the clearly defined services the facility is intended to provide. This is intended by government to transfer the risks associated with providing the service to those best able to manage them. This is an unnecessarily narrow approach to what is potentially a very effective way for the public sector to make best use of its constructed facilities.

Public–private partnerships should use strategic collaborative working guided by the seven pillars of partnering. This means that public authorities, consultants, contractors and specialists work together on the basis of cooperative teamwork to deliver the defined services the facility is intended to provide.

6.15 Construction as a Consumer Product Industry

A distinct form of strategic collaborative working is leading to the emergence of groups of consultants, contractors and specialists competent in the total construction cycle of development, production and use. They specialize in a specific type of building or infrastructure, provide sophisticated client support services and market the total package under a brand name.

Digital technology has revolutionized work techniques, systems, organization structures, supply chains and business processes. It enables modern industries to give clients guaranteed high quality, early delivery and low prices.

Construction is not insulated from these sweeping changes and groups of consultants, contractors and specialists have responded by developing distinctive ways of working that can be regarded as a third-generation of partnering. This brings together consultants, contractors and specialists able to deal with the total construction cycle of development, production and use as illustrated in Figure 6.2. The joint organization has all the knowledge, skills and resources to produce a specific type of building of infrastructure facility. Typical examples include apartment blocks, low-rise housing, student residences, warehouses, supermarkets, schools, hospitals, sports stadia and motorways. These products come with sophisticated client support services. The total package is marketed under a brand name to help provide potential clients with a clear image of what they can expect when they buy a new constructed facility. This in effect turns sectors of construction into a modern consumer product industry.

It takes time to build an organization able to deliver products and services to the best modern standards. It takes time to develop confidence in partners and learn how to cooperate. For this reason third-generation partnering organizations often develop from earlier experience with project partnering and strategic collaborative working with a major client. All the issues discussed earlier in this chapter apply to third-generation partnering. However, the emphasis is different because the level of investment and the nature of the risks involved are different. The main requirements for successful third-generation partnering go beyond the earlier parts of this chapter and are described in the following sections.

6.16 Governing Document

The basic agreement between the partners is a framework of non-binding principles set out in a governing document.

The strategic team should agree a governing document that sets out a framework of non-binding principles that describe how the clients, consultants, contractors and specialists involved in the arrangement will behave towards each other. It records the decisions made about all the issues covered by the seven pillars of partnering.

The strategic team members should ensure that the decisions are put into effect in their individual organizations. As a result all the partners should be able to confidently expect cooperative behaviour throughout the joint strategic collaborative working organization. This good effect is highly unlikely to be achieved by formal contracts between the partners, which tend to undermine strategic collaborative working. The best governing documents set down principles that are changed and developed on the basis of decisions and practice over years. They are a living guide not a formal contractual basis for claims or blame.

6.17 Organizing for Strategic Collaborative Working

Third-generation partnering organizations are diverse and the most highly developed become complex self-organizing networks that enable them to combine efficiency and innovation.

The form of organization that emerges to put third-generation partnering into effect results from a complex mix of the various elements described in Section 6.12 and Chapter 7, checklist 33. It exists alongside, and in part complements and in part conflicts with, the internal organizations of the individual partners. The resulting complexity is useful. Allowing many points of contact and overlap between the separate organizations increases the chances of finding answers to problems and responses to opportunities. Good ideas come from unexpected sources.

The result is that successful third-generation partnering organizations do not form neat and tidy structures of fixed roles and responsibilities. Instead they reflect the flexible attitudes and innovative thinking needed to respond to today's ever-changing challenges and demands. The organizations that develop over time develop into self-organizing networks that shape and reshape themselves in response to complex patterns of feedback and individual initiatives. The organizational features described in the following sections play important roles in these self-organizing networks.

6.18 Supply Chains

Third-generation partnering requires consultants, contractors and specialists to cooperate with their supply chains to improve performance.

Third-generation partnering provides time for construction to develop its supply chains. It is important to do this because most significant improvements in performance come from working with firms that form a supply chain. These significant improvements often grow from modest beginnings, perhaps aimed at improving quality or certainty of delivery. As these issues are discussed ideas for further improvements are identified. Problems are highlighted and solved. Processes previously carried out on construction sites are moved into factories. Design is integrated with manufacturing. Machines and robots replace human workers. Gradually, as confidence builds, bolder ideas are proposed and explored.

The results are often called lean production. Technology, quality, safety, efficiency and speed are radically improved to a stage where many people still wedded to the industry's traditional methods refuse to believe what is being achieved.

6.19 Communication and Learning Networks

Communication networks using digital technologies and face-to-face meetings develop into learning networks that search for new ways of improving performance.

Strategic collaborative working depends on wide communication using modern information and communication technologies and formal and informal face-to-face meetings. This should be developed into communication networks that support continuous effective communication between partners, clients, supply chains, competitors and regulators.

Communication networks should link senior managers and work teams because many ideas and much vital information come from the workface. They should aim to ensure that the interests of the joint organization are taken into account in key decisions in all the partners' organizations. They should help anticipate threats to the strategic collaborative working arrangement. Those from outside the organization are often relatively easy to identify. Threats can arise inside the organization particularly if one of the partners has to deal with a major change. It should be normal to discuss events that may cause any of the partners to revise their strategy or alter their priorities but commercial pressures may cause an organization to become secretive. Well-developed communication networks make this more difficult and so help protect the partnering arrangement.

Effective communication networks often develop into learning networks that bring external experts into the partnering arrangement to provide technical expertise, training, coaching and research skills. Such developments aim to encourage everyone involved to express ideas and views openly; support efficient work, innovation and real creativity; and develop a real commitment to the partnering organization's long-term success.

6.20 Design and Technology

Third-generation partnering sets challenging targets to encourage the development of new designs and innovative uses of technology.

The most important outcomes of third-generation partnering include new designs and innovative technological answers. They often come from value management studies undertaken outside individual projects. The procedures and techniques of value management are described in Chapter 7, checklist 29.

Dramatic improvements to quality, time and cost have been achieved by teams given challenging, often apparently ridiculous, targets. Typical of such challenges set by strategic teams are:

- Halve the cost of a supermarket whilst maintaining an image of quality.
- Construct a fast-food outlet in 24 hours.
- Reduce site labour by 80% by using prefabricated modules.
- Eliminate all site accidents.
- Achieve zero defects at every stage of construction.
- Provide the engineering services for a hospital as prefabricated modules commissioned off site.
- Reduce the client's labour force needed to run a new facility by 25%.

When a team is given time and resources to find answers to these challenging targets, wonderful things emerge. Whole elements are merged by finding new technologies. Work is moved from construction sites to factories, which can be based in regions where skilled labour is currently under-employed. Designs take account of manufacturing standards and so avoid unnecessary cutting and waste. Standard components are used rather than allowing unnecessary individual design. Mechanical systems are replaced by electronic devices. Design and manufacturing are merged. New construction techniques are devised. Construction begins to innovate with a purpose and channel the industry's creativity into efficiency and quality.

6.21 Ownership

Products developed by the partners should be owned jointly on a basis that allows them to be used inside or outside the partnering arrangement.

Ownership of things developed by the strategic collaborative working arrangement can raise difficult issues. It may be that one organization makes a large investment that delivers much higher returns inside the partnering arrangement than could be obtained outside. They are therefore at risk if the arrangement ends. Equally problems come from developments that can be exploited better outside the arrangement.

These kinds of problems centre on ownership of innovations. Ownership should be agreed so as to avoid causing any sense of exploitation or creating temptations for partners to defect.

A practical approach is to agree that innovations should be owned jointly and can be used by the partners inside or outside the strategic collaborative working arrangement. Outcomes in terms of the allocation of benefits and costs are determined by the market. This means that where an innovation is best used inside the partnering organization, partners have an incentive to allocate costs and benefits in ways that ensure this happens. Equally if some partners have strengths that enable them to exploit an innovation in a new market, they can do so as long as this does not inhibit the joint ownership. These arrangements depend, like much else in strategic collaborative working, on partners being open with each other and agreeing the best courses of action so there are no big surprises.

6.22 Ambiguity and Balance

Partners need to tolerate contradictory answers being adopted in the short term in different parts of the overall organization. This encourages efficiency and provides the flexibility needed to innovate, deal with change and survive long term.

Teams involved in strategic collaborative working arrangements need to recognize that their decisions often require them to balance different interests. They may be faced with choices between internal and external, individual and group, efficiency and creativity, planning and flexibility, control and initiative, and cooperation and competition.

Teams need to accept that there will be some ambiguities and contradictions in the short-term answers adopted in different parts of the joint organization. This should be accepted because organizations that tolerate a variety of behaviour around a central core of focused work tend to survive long term. Tolerant flexible organizations often respond to major changes by using answers developed independently by a team exploring its own individual ideas. This flexibility gives the team decisive advantages over organizations that achieve superefficient consistency. Focused superefficiency may win out in the short term but by continuing to drive forward in a straight line even when the road changes direction, the short-term winners tend to crash.

One of the great advantages of strategic collaborative working is that it allows joint organizations to cope with the ambiguities involved in balancing efficient consistency with tolerance and flexibility.

Flexible networked organizations make it relatively easy to tap into new knowledge and skills. New partners, external experts and research organizations can be brought into the organization and begin effective work quickly when cooperation is the established normal way of working.

The great ambiguity inherent in strategic collaborative working is that cooperation makes strategic collaborative working organizations more competitive. They can deliver more for their clients. They have all the benefits of lower costs, faster production and more reliable quality. They have access to wider knowledge bases so they recognize, understand and respond to major trends earlier. They can afford to invest more in developing new technologies and designs than competitors who are not using mature strategic collaborative working. They compete on the basis of high-quality processes, products and services, which are what modern clients expect in all aspects of their own businesses.

6.23 Marketing

Third-generation partnering depends on totally professional marketing to create long-term sustainable businesses by building client loyalty.

Successful industries use marketing to deliberately foster demanding clients. Professional marketing is an essential part of building successful long-term businesses. It provides a robust basis for packages of products and services tailored to the needs of distinct categories of clients. It helps if distinctive labels are established for the packages such as luxury, executive, high-tech, family and economy. The labels linked to distinctive brand images help clients understand the choices available. As a result demand for the best-value products and services increases so that consultants, contractors and specialists delivering consistently poor value are forced out of business.

Construction is subject to these same pressures. Many consultants, contractors and specialists already have experienced, demanding clients who do their own research to establish what products and services they can expect to get and on what terms. Increasingly there is good information available to clients from industry-wide organizations and some individual consultants, contractors and specialists. More generally television programmes that help everyone realize what they should expect from new houses or apartments and provide detailed information on prices are increasingly popular.

All this is moving in the right direction and clearly indicates the need for good marketing. It is vital that consultants, contractors and specialists respond. An important reason is that successful strategic collaborative working depends on professional market research. The consultants, contractors and specialists involved need to base their decisions on up-to-date information about the factors clients take into account in deciding whether or not to go ahead with construction projects. The aim is to make it easier to buy and own new buildings or infrastructure. Consultants, contractors and specialists that get their marketing right will increase their market share.

Marketing is much more than market research. It involves working closely with clients, understanding their real needs, making it easy for them to understand the choices, and delivering on every explicit and implicit promise. It does not mean expensive advertising campaigns or huge sales forces. It is much more about providing clear information that is easily available to potential clients. It is the final piece of the jigsaw that will turn construction into a modern consumer product industry.

BAA Lynton roll out mark 3

Case Study Reference: M4i 101

A team led by BAA Lynton delivered a prestigious £9m, 77,000 sq. ft office in record time. The air-conditioned four-storey building at Stansted airport provides KLM's new headquarters and is designed to be sublet in quarter floors if necessary. BAA nominated the building as an M4i demonstration project.

This is the third version of an office product. LMK's Construction Manager, Bob Williams explained: 'With the team, we developed this as a "product" building that can be placed in virtually any location to meet the desires of individual clients.' Asked if he would use the product when a unique architectural statement is required, Williams replied: 'No, but you would have to pay a lot more and wait much longer for a new design. In practice there are few projects that demand a truly unique solution.'

Alistair Taylor of BAA is delighted with the result, which comes from BAA taking an enlightened view in encouraging innovation. They have a framework agreement with a shortlist of suppliers. This encourages the development of designs and technologies at lower prices than ever result from competitive tenders. Architect Peter Runacres of REID architecture said: 'The project demonstrates that money spent on design development, with early inputs from all the key suppliers and contractors, improves the product before start on site and benefits the project in the long term.'

Alistair Taylor of BAA said: 'Time is critical. By adopting the office product, we saved at least three months in the time from planning application to starting on site.'

LMK's Construction Manager, Bob Williams, reckons on saving £50,000 for every week saved on site: 'Above ground it took us five weeks less than the previous version, saving about £250,000.'

Reduced project time allows early release of areas to tenants, a big plus for the office developer.

Value for money is high on BAA's shopping list. For the quality of office building provided for KLM, the normal construction cost is £80/sq. ft. The latest model is 4% lower at £77/sq. ft.

LMK finds that building the office product is easier than bespoke offices. An individually designed office building would have needed at least five construction management staff on site but only four are needed to build the office product.

Working with a known building product has a beneficial effect on safety on site. The accident frequency rate was nearly 50% better than industry standards set by the Health and Safety Executive.

Roles reversed in successful PFI school

Case Study Reference: M4i 150

When the Private Finance Initiative was conceived, construction's response was to put contractors in the driving seat, but Building Design Partnership has bucked this trend with their wholly owned Private Finance Initiative Company, Campus Projects. The benefits of designers taking on Private Finance Initiative projects is demonstrated by Campus Projects' stunningly successful Drumglass High School at Dungannon in Northern Ireland. The £6m school for 500 pupils was designed and built in just 14 months.

The project provided a showcase for what can be achieved when designers take the initiative in partnering with construction firms. Campus Projects' school outperformed its peers in respect of almost all criteria set by the local education authority. As a bonus the buildings are pre-wired for the next generation of IT facilities to be installed under the Classroom 2000 initiative.

Campus Projects now has a tested product ready for marketing to other education sector clients. This is an excellent example of delivering product families, as recommended by the Egan Report. The main benefits included the following:

- Capital cost – the assessment of school costs included comparing the net present worth of the predicted 'whole life' costs against the Public Sector Comparator (the model of equivalent costs under traditional procurement). This clearly demonstrated that the school authority got good value for money.
- Construction time – the Drumglass project was completed at least one year ahead of other projects launched at the same time. Buildability achieved by working closely with the contractors was critical in achieving this remarkable speed of delivery. Principal teacher Derek Wilson said: 'It's amazing what was achieved in 14 months.'
- Turnover and profit – this was Building Design Partnership's first experience as a Private Finance Initiative contractor and the financial results encouraged them to develop their supply chains to deliver more projects. Similar positive outcomes motivate William Martin, Managing Director of contractor H&J Martin, in expressing his respect for Campus Projects' leadership and his eagerness to work again with the same designers.

Using design and construct to drive innovation

Case Study Reference: 035

London Underground has established a group of preferred contractors and used partnering to build up its expertise in railway earth structures. They adopted a 'design and construct' contractual framework to give the contractors total responsibility for projects.

London Underground actively encouraged innovation. By using both conventional and innovative engineering techniques, the contractors achieved remarkable improvements in performance. Key benefits from this partnering initiative included:

- Cost of embankment stabilization reduced by up to 50% in five years.
- Customer service improvements.
- Earth structure-related speed restrictions reduced from 14 in 1995 to zero in 2000.
- Asset information systematically updated and improved.
- Development of new engineering techniques.
- Reduced maintenance expenditures through improved track quality.
- More strategic, long-term approach to the management of London Underground's assets.

T5 Buy Club – How M&E contractors pool purchasing at Heathrow Terminal 5

Case Study Reference: BAA Heathrow

Heathrow Terminal 5's M&E Buy Club ushers in a new era of openness, collaboration and striving for 'world class' results. It replaces the industry's traditional approach based on secretive deals with favoured suppliers. Instead, the Buy Club pools the expertise and buying power of five first-tier mechanical and electrical contractors and arranges for each of 13 specializations to be supplied from (generally) one supplier who is then responsible for supplying all 16 projects at Terminal 5.

Looking beyond Terminal 5, the Buy Club model can be either adopted entirely or adapted to suit the circumstances facing other projects. The only essential prerequisite is having more than one specialist contractor doing similar work.

As a direct result of this partnering approach, BAA made savings of between 10 and 30% of the budget for mechanical and electrical equipment and materials. Commodities such as cables and bulk supplies are yielding savings at the lower end of this range while inputs, which are more design intensive, such as low voltage switchgear, are producing savings of close to 30%.

Instead of asking potential suppliers to do up-front work as a favour, the Buy Club ensures their commitment by appointing them early and engaging them in design. The results are 'lean' manufacturing and installation. Collaboration to meet Buy Club targets requires planning by first-tier suppliers. One important benefit is that for all but the first project, mechanical and electrical procurement is taken off the critical path.

Early agreement of benchmark prototypes means suppliers are actively involved in design, know exactly what is required and can budget and plan accordingly. An open-book approach to quality reveals issues before they become problems.

The Buy Club has been so successful for mechanical and electrical contractors that BAA is using the same approach for Terminal 5's £200m fit out and £50m communication systems packages. The model of having first-tier contractors and their suppliers partnering to drive down costs is applicable to other large projects as well as programmes of smaller projects.

Techniques and checklists

This chapter provides a detailed manual for practitioners using partnering for the first time, and for those developing their partnering skills and knowledge. It is equivalent to the appendices in the *Code of Practice for Project Management for Construction and Development*. It does not duplicate the descriptions in the sister publication; rather, it describes techniques particularly associated with partnering and provides checklists for clients, consultants, contractors and specialists to help ensure that they are using best practice.

Where the specific constructed facility produced by a project is referred to, the term 'building' is used. Clients involved with infrastructure projects need to alter 'building' to 'infrastructure' or to a specific type of infrastructure. Also where the facility is referred to as a new building, this includes refurbishment, alterations and maintenance work.

Early Decisions about Projects and Partnering

Establishing Project Partnering

Developing Project Partnering Organizations

Project Systems for Partnering

Management Techniques in Partnering

1 Client's Checklist of Main Questions

This checklist provides questions and identifies issues that clients should consider before making firm decisions about a construction project. Some of them may require the client to obtain expert advice.

Why is a new building needed?

The client should write a statement describing why a new building is needed and the benefits it must deliver. This should describe the client's organization and the way it works, and how the proposed building contributes to the organization's activities. The statement provides an outline business case for the project and defines the client's criteria for determining whether it is successful.

Is a new building essential?

- Is there an existing building that provides the spaces needed that can be bought or rented?
- Can the organization's activities be reorganized to allow them to be carried out in the organization's existing buildings?

Where is the new building needed?

- Is there an existing building that could be altered to provide the spaces needed?
- If an existing building is available, does it impose significant physical or planning constraints?
- Does the client have a site in mind for the building?

What kind of new space is needed?

- What spaces are needed?
- Who needs to come into the building and which spaces will they need to go into?
- What physical things come into or go out of the building and which spaces are involved?
- How does the building contribute to work processes and productivity?
- Which spaces need to be near each other?
- What or who is to be accommodated in each space?
- What activities will take place in each space?
- What services need to be available in each space?
- What internal conditions need to be provided in each space?
- Is sophisticated control over internal comfort conditions needed or is natural light and ventilation acceptable where it is suitable?
- What level of quality is needed in each space?
- What level of security is needed in each space?

- Do the spaces need to be flexible to allow for different uses in the future?
- Does the building need to allow for future expansion?

What kind of building does the client want?

It often helps for the client to look at buildings that provide spaces similar to the one they have in mind; and at buildings near the proposed site for the new building.

- Do any of these provide the style and feel needed?
- In what ways should the building be different from those looked at?
- Will an ordinary building that provides good value at a low cost provide what is needed?
- Will a piece of architecture that makes a bold statement about the style and importance of the client's organization provide what is needed even if it is expensive?
- Does the building need to reflect or help shape the organization's culture in some way?
- Are there specific design standards or other aesthetic considerations that need to be taken into account?
- Are there specific quality standards for design, materials, components and workmanship that apply?
- Are there any aspects of the building's running or maintenance that need to be taken into account?
- Is it important that the building is suitable for other uses so that it is intrinsically valuable and can be sold at a profit if this became necessary in the future?
- How long should the building be designed to last?

Does the client want to influence the building's impact on the environment?

- Does the client want consultants, contractors and specialists to work in their normal way within the regulations and let the impact of their new building be what ever it turns out to be?
- Does the client want to get expert advice on aspects of the building's impact on the environment?
- Does the client want to influence the way the building changes neighbouring streets and public spaces?
- Does the client want to set limits to emissions of pollution during construction?
- Does the client want to set limits to emissions of pollution from the building when it is in use by their organization?
- Does the client want to set limits for the building's energy use?
- Does the client want the building to have specific effects on the local community?

- Does the client want the building to have specific effects on the local economy?
- Will the local authority impose specific environmental conditions on the building?

How much can the client afford to pay for the new building?

- Could a new building enable the client's organization to make its operations more effective, efficient and reliable?
- Will the new building allow the client to expand their business?
- Will it allow the client to offer better products or services to their customers?
- Will it allow the client to operate more efficiently?
- Will it help the client to recruit or retain staff?
- Will it allow better communication inside the client's organization or with suppliers or customers?
- Will the work flow be streamlined?
- Will it allow the client to reduce the costs of running and maintaining the space their organization occupies?
- What is the overall value to the client's organization of the benefits identified by the above questions?

How will the client pay for the new building?

- Will the project be paid for from the client organization's own funds?
- Does the client need to borrow the money and if so what information needs to be made available for financial institutions?
- Does the funding need to consider capital, running and maintenance costs?
- Are there any cash-flow constraints?
- Does the client need the construction industry to organize or provide the finance?

How quickly is the new space needed?

- What is the latest date the new building can be finished?
- How critical is the timescale?
- What are the advantages of an earlier finish?
- Does the client need the new building to be totally complete with everything working properly when they move in or can they accept the builders having some work to finish?
- Does the client have specific requirements for the timing or sequence of construction activities?
- Are there key decision points that must be met?

Does the client want to outsource facilities management services for the new building?

- Does the client want an external organization to provide facilities management services for the new building?
- Which facilities management services does the client want to outsource?
- For what length of time does the client want facilities management services to be provided?
- How does the client want the facilities management services to be paid for?
- Does the client want the payment structure to include rewards for improved performance?
- What criteria will be used to monitor the quality of facilities management services?
- Does the client want the organization providing facilities management services to have some responsibility for the new building's contribution to the client's business objectives?

How much does the client want to be involved in the building project?

- Does the client want to agree a clear statement of the building they need and then leave it up to consultants, contractors and specialists to deliver exactly that?
- Does the client want consultants, contractors and specialists to carry the risks involved in the new building or are they prepared to discuss each of the risks and then decide the best way of dealing with them?
- Does the client want to commission studies of the various ways the new building can be used to ensure that it delivers the maximum possible contribution to the client organization's efficiency?
- Does the client want to get involved in design decisions so they understand all the options and then choose what goes into the new building?
- Does the client want to be involved in decisions that will influence the costs of cleaning and maintaining the building?
- Does the client want to be involved in detailed cost planning to ensure they get the best possible value throughout the building?
- Does the client want regular progress reports?
- Does the client want to influence the approach to health, safety and welfare in construction, operation and maintenance?
- Does the client want to know when problems arise?
- Does the client's organization have formal approval processes that the project team will have to work to?
- Does the client want to be involved in planning the handover of the finished building to their organization?

Who will be involved in the building project?

- Who will represent the client's organization and what are their responsibilities and interests?
- What external organizations have an interest in the project?
- What discussions have taken place with the local planning and other regulatory bodies?

Answering these questions may cause the client to decide that there is no need for a new building. Alternatively the answers may confirm that a new building appears likely to provide benefits for the client. In this case the client should organize the production of a written statement of the answers to provide the initial brief for the project sponsor.

2 Choosing a Standardized Solution or an Original Design

This checklist provides questions that help clients decide whether a specific standardized solution meets their objectives or whether they need an original design.

- What benefits will the standardized solution provide?
- Will the standardized solution help reduce the costs of the client organization's activities?
- Will the client be able to provide a better service to their customers?
- Will the standardized solution make it easier for the client to recruit and retain staff?
- What levels of quality and comfort will be provided?
- When will the client's organization be able to move in?
- How certain is that date?
- How much disruption will there be to the client organization's activities?
- What will the standardized solution cost in total?
- How certain is that cost?
- What is the best way to finance this investment?
- How do the running and maintenance costs compare with current costs?
- How certain are those costs?
- What do the people who will use the building think of the standardized solution?
- What does the local authority think of the standardized solution?
- What do the neighbours think of the standardized solution?
- Are there any groups of people who will object to the standardized solution and if so what should the client do about them?
- What are the biggest risks involved in going ahead with the standardized solution?
- What are the biggest risks involved in not going ahead with the standardized solution?
- What is the value of the new building if it were sold as an empty building?

If the client is happy with the answers, they should seriously consider using the standardized solution. If no satisfactory standardized solution can be found, the client should consider commissioning an original design.

3 Choosing the Form of Project Organization

These checklists will help clients unfamiliar with the main forms of project organization used by the construction industry decide whether a particular option is suitable for them and their project.

Clients should use the guidance provided in Chapter 6 if they want to establish a strategic relationship with consultants, contractors and specialists that goes beyond an individual project, whether by means of a public–private partnership, which includes private finance initiatives, or some form of build, operate, transfer arrangement. Any of the various forms of project organization dealt with in this checklist can be used within those higher-level arrangements.

General contracting

- General contracting is the construction industry's traditional approach and is based on long-established professional and craft roles. It is suitable for clients that want an individual design of high aesthetic quality and are prepared to spend time selecting and working with a number of consultants, contractors and specialists.
- The client employs consultants including architects to design the overall building and all its details; engineers to design the structural elements and engineered services, and check that the building is safe and comfortable; and quantity surveyors to check that the designs can be afforded within the client's budget.
- In addition specialist contractors may be employed as part of the design team because their expertise is needed for the design of key elements or systems of the building.
- The design information is used in selecting a main contractor to construct the building.
- Most of the direct construction work is undertaken by specialist contractors employed by the main contractor as subcontractors.
- Construction work is supervised by the consultants to ensure that the design is correctly interpreted, to undertake quality control, to authorize payments and to deal with any claims from contractors for additional time or money.
- The traditional general contractor approach has weaknesses:
    - It is essentially a sequential process that is inherently slow.
    - It separates design from first-hand experience of construction and so tends to produce designs that are difficult and therefore expensive to construct.
    - It tends to combine new technologies, unfamiliar construction details and a fixed price established by competition making conflict and claims inevitable.
    - The quality of traditionally produced buildings can be patchy with too many defects when the building is handed over to the client, not all of which can be put right.

    - It produces conflicts between consultants, contractors and specialists, which all too easily result in clients being faced with demands for extra time and money.
- These weaknesses are the main reason why alternatives, notably design build, prime contracting and management approaches, have been devised.

Design build

- Design build is suitable for buildings that use well-established designs and standard materials and components.
- Design build works best if the client is able to produce a clear and certain description of the building they want, what they are willing to pay for it and when they need it.
- For all except the simplest of buildings, it is likely that the client will need to employ consultants to help produce a description of the required building capable of providing a sound basis for a design build contract.
- In using design build the client enters into a contract with a contractor who takes responsibility for doing everything necessary to produce the building the client wants for an agreed sum by an agreed date.
- The client's internal team must establish controls to ensure that what is being designed and produced is what the client wants. This normally means employing consultants to advise on the implications of the contractor's design and construction proposals, undertake quality control, advise on costs, check requests for payments, advise on changes to the completion date and agree the final account.
- The client has to decide how much information they need about their new building before they will enter into a contract with a design build contractor.

Minimal statement

- The simplest approach is to produce a list of the activities to be accommodated and the functions to be performed by the building with little or no design or specification of the actual building.
- This approach gives the contractor control of the whole design process and usually results in a competently designed building and the project being completed on time and within budget.
- This approach makes the fewest demands on the client's time but may mean they miss opportunities for the building to provide more than they thought of initially.

Outline design

- Some clients want to consider several design options and employ consultants to discuss their requirements and produce design concepts for buildings with different layouts of spaces, different external and internal appearances, and different positions on the site.
- The approach gives the client the opportunity of considering the main design options and selecting the most suitable.
- The consultants produce an outline description of the selected design in a form suitable for a design build contract that usually comprises plans and elevations of the required building supported by the functional requirements of the main elements and systems, plus a performance specification or an indicative technical specification that defines the quality required.
- This approach produces variable quality buildings and often leads to the contractor claiming extra time and money.
- The best outcomes from this approach result when the contractor is given responsibility for completing the design and producing the required building.
- The worst outcomes result from consultants producing an outline design that needs considerable further development by them after the design build contractor is appointed. Even when the employment of the consultants is taken over by the contractor, there is a sense of divided responsibilities, which lead to disputes and poor performance.

Detail design

- Some clients want to exercise control over all aspects of the design so they can be sure they get the building they need. They employ consultants to produce a complete design of the building and all its details and a specification of the quality of each part and the performance of the overall building.
- This approach gives the client opportunities to consider every aspect of their building but puts cost and time at risk until the design is agreed.
- A contractor is selected and given time to check the design information, carry out tests or checks on site, raise questions and suggest improvements to the design.
- The agreed design information is used as the basis for a design build contract in which the contractor takes responsibility for producing the building.
- This approach creates a single point of responsibility for producing the building and tends to result in the project being completed on time and within budget.

Prime contracting

- The prime contracting approach is suitable for clients that want an original design and also want a single firm to take responsibility for producing their new building from an early stage in the project.
- A firm is appointed to provide overall leadership of the construction inputs needed to produce a new building that gives the client value for money.
- The prime contractor may have a background in consulting or contracting. The key is that they are competent to provide leadership for an integrated project team using cooperative teamwork to deliver excellent value for money by producing the required building reliably and efficiently.
- Essentially the prime contractor is responsible for leading the project team in:
  - Working with the client's internal team to produce a statement of the client's objectives for the new building.
  - Producing design concepts consistent with the client's objectives.
  - Ensuring that the design development and detailing are consistent with the selected design concept and the client's objectives.
  - Developing a construction strategy consistent with the client's objectives.
  - Developing detailed plans for the manufacturing and construction activities and ensuring they are put into effect efficiently and reliably.
  - Developing detailed operating and maintenance methods and techniques.
- In cases where the client wants to buy facilities management services, the prime contractor may have a continuing responsibility for these services.
- Consultants and specialists needed to carry out all these tasks are brought into the project team as soon as their role is identified and a suitable firm selected.
- The commercial arrangements with each consultant and specialist are agreed by the prime contractor within an overall framework formed by the client's objectives, the business case for the project and the required completion date.
- This approach provides clients with the opportunity to influence decisions and the outcomes tend to reflect the certainty and clarity of their objectives. Decisive clients tend to get a good-quality building on time and within budget. Clients that change their requirements and delay decisions may well get a mediocre building late and over budget.

Management approaches

- The management approaches are suitable for clients that want an original design for their building constructed efficiently within the time and costs needed to fit the business case.
- There are two main forms of the management approach. The first is management contracting in which the client employs a design team of consultant architects and engineers and a management team. The management contractor enters into subcontracts with specialist contractors.
- The second is construction management in which the client employs a design team of consultant architects and engineers, a consultant construction management team and specialist contractors.
- The management approaches are used by many experienced clients because they want to be closely involved in the key decisions made by their project teams. Indeed experienced clients often chair project team meetings to ensure that their interests are taken into account.
- It is not sensible for an inexperienced client to attempt to play the central role adopted by experienced clients. Where an inexperienced client wants to use a management approach, they can appoint a consultant project manager as the project sponsor to undertake this central role.
- Management contracting and construction management have slightly different strengths but when used properly are substantially identical.
- Construction management involves the client in forming contracts directly with the specialist contractors, while with management contracting, the specialist contractors are employed by the management contractor as subcontractors.
- Management contracting creates a tendency for the firm providing construction management to be regarded as a contractor and specialist contractors as subcontractors. These perceptions can inhibit design consultants in allowing them to make a full contribution to the design.
- Construction management creates a clear consultant role for the construction manager and encourages specialist contractors to be committed to the client's objectives and contribute fully to the design because they have direct contracts with the client.
- The differences can be rendered insignificant from the client's point of view by experienced firms but the differences exist and clients should check exactly what liabilities and risks are involved in each of the contracts they enter into.
- Both management approaches can result in buildings of variable quality that in most cases are produced reliably on time and within budget.

Key issues

- The construction industry has made considerable progress in improving its performance in recent years. The following features of best practice can be used with any of the forms of project organization and clients should check whether it is appropriate for their chosen approach to include the following.

Quality assurance

- The construction industry has made considerable progress in adopting good quality assurance systems. The client's internal team should ensure that the quality assurance systems used by all the consultants, contractors and specialists they employ are checked against the requirements of good practice described in checklist 24.
- It is realistic to ask project teams to aim for zero defects in new buildings when they are handed over to the client. The client's internal team should discuss this aim with the project team and not accept any argument that good quality costs more. The costs associated with correcting defects far outweigh the effort needed to do work right first time.
- An effective way of reinforcing quality is to make payments to contractors dependent on quality controls being properly carried out.

Complete design before construction begins

- An effective way of removing many of the inherent problems of requiring contractors to work to designs produced by independent consultants is for the design to be complete before the contractor starts construction work. This can be reinforced by the consultants having no authority to change the design.
- Some experienced clients who use this approach require the design consultants to sign the design information as complete and suitable for the main contractor to build from before construction begins.

Use information and communication technology

- Construction processes can be speeded up and made more reliable by using ICT.
- ICT can reduce the time taken for information to travel between individual work teams and reduce the risks of working on information that is out of date.
- ICT makes it easier for construction projects to be programmed effectively, work teams to know when to do their work and progress records to be kept up to date.
- ICT can speed up the work of individual work teams by automating routine information processing.

- These benefits depend on work teams being experienced at using compatible systems and the client's internal team should take this into account in selecting consultants, contractors and specialists to provide work teams.

Coordinated project information

- The construction industry has developed a scheme for coordinating the design information that independent consultants produce so the design information produced by each separate consultant is consistent with that produced by all the others, and that in total the design information is complete. It is sensible to employ consultants who use this scheme, which is called coordinated project information or CPI.

A common project office

- An effective way of speeding up the design process is for key consultants to work in a common project office.
- Before setting up a common project office it is important to do a preliminary risk assessment to ensure as far as possible that the project will not suffer a major delay since this would almost certainly lead to abortive costs.
- It is an advantage for consultants to be experienced in working together in a common project office. People have to adjust to working in close cooperation with other professions and it may not make sense for this learning process to take place on the project.
- There are other costs associated with a common project office including the direct costs of the office, and travel and other costs for people working away from their normal office base.
- Despite the direct costs, the greater efficiency provided by working in a common project office means consultants' costs should be no higher than when they work in their own offices.
- There are likely to be considerable benefits for the whole project team from the resulting faster, more certain, more innovative work. The results can be dramatic improvements in the speed and effectiveness of communication. Design times can be cut from many months to just a few weeks of intensive and effective work.

Employ a project manager

- Employing a project manager to coordinate the work of all the separate consultants, contractors and specialists can result in greatly improved cost and time performance both in terms of efficiency and certainty of delivery.
- Many traditional consultants feel uncomfortable working with a project manager. They complain that they are not allowed the time or resources needed to explore enough design options, especially during the detail design stages, and that the resulting buildings are not as good as they could and should be.
- If the client decides to employ a project manager, the above issues should be discussed with potential project managers and design consultants before making a final selection.

Partnering

- Partnering requires the construction industry's traditional emphasis on independent work to be abandoned and replaced by cooperative teamwork.
- The general contractor approach does not establish a complete project team until after the design is complete and the main contractor is selected. This inevitably means that partnering has less potential to provide benefits than when a design build, prime contracting or management approach is used.

4 Choosing to Use Partnering

This checklist will help clients, consultants, contractors and specialists decide whether to use partnering. To be sure that partnering will suit them, they need to be able to answer 'Yes' to the following questions on behalf of themselves and their representatives.

- Do you agree that working in cooperation with the other members of the project team will give you more benefits than if everyone concentrates on looking after their own interests?
- Do you want the firms you employ to make a fair profit?
- Are you willing to spend time ensuring that partnering is successful?
- Are you prepared to be open about your organization's interests in searching for mutual objectives with the other members of the project team?
- Are you prepared to be questioned by the other members of the project team that justifiably expect full and open answers?
- Are you willing to join in a cooperative search for solutions immediately problems arise without trying to allocate blame to individuals or firms?
- Are you willing to make decisions quickly for the good of the project?
- Are you willing to finance the preparatory stages involved in partnering?
- Are you willing to change your internal procedures if they inhibit partnering?
- Are you willing to spend time devising performance measures that reflect agreed ways of working and planned outputs?
- Are you willing to replace your representatives if the rest of the project team decides they are not acting in a manner consistent with partnering?

The client should be able to answer 'Yes' to at least some of the following questions because they define the benefits the client wants from partnering. If the client does not want any of these benefits, there is no point in partnering.

- Are you prepared to work with the project team to find designs that will increase the benefits the new building provides for your organization?
- Are you prepared to work with the project team to find ways of making sure the project meets its budget and completion date?
- Are you prepared to work with the project team to find ways of making sure the new building is fully complete with zero defects when it is handed over?

- Are you prepared to work with the project team to find ways of producing the building more quickly than normal?
- Are you prepared to work with the project team to find ways of producing the building at less than the normal cost?
- Are you prepared to work with the project team to find ways of providing other benefits for your organization?

5 Senior Managers' Partnering Checklist

This checklist will help senior managers ensure that their organization is using partnering effectively. It identifies issues that should be given attention by teams using partnering.

Basic checks on successful partnering

- Everyone involved in partnering should understand that it is a set of practical actions that deliver benefits when they are applied steadily and consistently on the basis of commitment and hard work.
- Partnering should achieve significant benefits even when it is being used for the first time.
- The level of benefits should increase as teams work together on a number of projects.

Key requirements for cooperative teamwork

- Invest in training and workshops when teams have not worked together before or are new to partnering.
- Create conditions that encourage and reward cooperative behaviour.
- People should take account of others' interests and understand that this concern is in their own best interests.

Developing successful project partnering

- Invest in developing ever more effective relationships between work teams.
- Provide continuity by forming project teams from work teams that have established efficient relationships.

Avoiding problems

- Be aware of potential conflicts of interest where partnering firms are partnering with competitors, but maintain the open communication required by partnering and build in checks to ensure that commercial information is not abused.

Ensuring effective decision-making

- Partnering teams should discuss widely before decisions are made so they consider a range of points of view and accept that good ideas often come from unexpected sources.
- The interests of all the stakeholders should be taken into account by devising ways of ensuring they are consulted and their interests taken into account.
- Use task forces that include external experts to help ensure that divergent but useful ideas are taken into account.

- If a talented individual capable of making a significant contribution to a project is unwilling or incapable of cooperative teamwork, consider creating space within the project team for them to work independently. Give them opportunities to comment on team decisions but do not waste resources trying to include them in partnering activities.

Dealing with weaker partners

- Recognize that even powerful organizations using partnering benefit from producing the maximum net benefits and sharing them in a manner that is sufficiently fair to motivate everyone to do their best possible work.
- Be prepared to help weaker partners; for example, through training in the actions needed to meet exacting standards, and providing advice on finance, control systems or other key organizational issues.

Setting targets

- Set targets that are challenging and achievable. Initially they can aim at substantial improvements in one specific area. Then broader and tougher targets can be set as success grows.
- Take the time and care needed to ensure that targets take account of the interests of everyone involved with a construction project.
- Ensure that targets are easily measured to avoid arguments about whether improvements have been achieved.
- Ensure that targets can be achieved without damaging those further down the supply chain because the benefits will be small in comparison with what could be achieved by fully involving subcontractors, suppliers and manufacturers.
- In setting targets partnering teams must also establish a firm datum of established performance that must be achieved so that teams do not neglect established good practice.

Ensuring continuous improvements

- Establish good standards and procedures and use them unless they conflict with agreed mutual objectives.
- Use task forces to develop robust improvements to standards and procedures.

Establishing feedback systems

- Invest in setting up and using good feedback systems that guide partnering teams in searching for ways of improving their performance.
- Provide significant incentives for project team members at the end of each project to collate and feed back their experience.
- Ensure that feedback is produced and used by people carrying out direct design and construction work as an

integral part of their normal work so it has immediate relevance.

- Involve work teams in direct face-to-face discussions when their own good ideas and experience are developed for use by other teams because they understand the details and potential problems.
- Invest in feedback from project to project to provide a powerful spur to long-term substantial improvements in performance.
- Ensure that senior managers have strategic feedback about the costs and benefits of partnering needed in guiding their businesses.
- Senior managers should visit the workface regularly and talk to staff about their work including their use of feedback so important information is never lost in badly designed feedback systems.

Strengthening partnering firms

- In introducing partnering into a traditional organization, provide support for a strong individual in taking risks to establish cooperative behaviour in the face of any opposition from colleagues convinced of the benefits of competition.
- Senior managers need to give partnering teams the authority, time and resources to decide their own best ways of working. The support needs to be maintained when initial costs are incurred before the benefits emerge.
- Senior managers need to accept that new partnering teams have to learn how to work together and some initial decisions may be poor and have to be changed.
- Senior managers should not take over but act as a coach or mentor, provide training and encourage partnering teams to find effective ways of working.
- Accept that partnering changes the nature of work for many people in requiring more face-to-face contacts and this may require training and organizational changes.
- Ensure that everyone involved in partnering has sufficient authority to make decisions that will be supported by their own organization.

Developing successful strategic arrangements

- Ensure as far as possible that strategic arrangements provide partners with a volume of business that makes commercial sense to them.
- Encourage 'competing' suppliers to cooperate in finding ever more effective answers they can all provide for the buying organization.
- In developing a strategic relationship look for opportunities to develop more profitable new businesses.

Maintaining partnering arrangements

- Work on the basis that partnering arrangements will continue long term to give everyone involved the confidence to put their best efforts into ensuring that the partnering arrangement is successful.
- Recognize that circumstances change and a successful arrangement may become less attractive to some partners. If this happens treat firms that leave generously.
- Recognize that the long-term interests of a strategic arrangement may require some aspects of a specific project to be compromised in order to test a major innovation or new design concept.

6 Evaluation Sheet for Selecting Firms

This technical description provides an example of an evaluation sheet that illustrates the way performance is balanced against price.

The weighting is agreed by the client's internal team before questionnaires or tenders are invited. The scores are based on detailed questions relating to the main criteria agreed by the client's internal team before questionnaires or tenders are invited. In the example below the scores are awarded out of 100. The performance threshold is established by the client's internal team to ensure a competent firm is selected. In the example below the performance threshold is set at 300 out of a possible total of 400. A firm failing to achieve the threshold should not be considered. The scores are awarded by each assessor individually and discussed by the selection team until a consensus is reached.

Performance weighting	80%	Project:	Winterbourne School
Price weighting	20%	Firm:	Hamilton Design Studio
Performance threshold	300	Assessor:	Ray Smith

	Project weighting	**Score awarded**	**Weighted score**
Performance Criteria			
Firm's track record	15%	80	12
Technical competence	20%	75	15
Project organization	15%	60	9
Teamwork	30%	85	25.5
Performance Total		**300**	
Price Criteria			
Price level	15%	60	9
Price certainty	5%	90	4.5
Total Weighted Score			**75**

The following list will help the client's internal team select evaluation criteria.

Firm's track record

- Internal culture and organization
- Flair, commitment and enthusiasm
- Environmental concerns and systems
- Financial status and resources
- Physical resources
- Workload

- Relevant experience on similar projects
- Local experience.

Human resources

- Management skills and knowledge
- Management systems
- Technical skills and knowledge
- Equality and diversity in human resource policies.

Client relations

- Client communication systems
- Understanding the client's objectives
- Compatibility with client and other project team members.

Project team player

- Project team communication systems
- Dispute avoidance and resolution techniques
- Risk management skills and techniques
- Supply chain involvement
- Resources allocated to the project.

Performance controls

- Quality control systems and track record
- Time control systems and track record
- Health and safety systems and track record
- Cost control systems and track record.

Value and price

- Value for money track record
- Adding value through innovation
- Value management skills and techniques
- Whole life costing skills and techniques
- Price level for the project
- Price certainty for the project
- Value for money for the project.

7 Competitive Tenders in Partnering

Traditional competitive tenders aiming at establishing the lowest price for a given design are incompatible with partnering. When competitive tenders have to be used the aims should be to select firms able to use cooperative teamwork with the rest of the project team and establish the basis for calculating the price for their work.

Public sector clients often want to use competitive tenders. They should do so in ways that enable them to use partnering strategically with the aim of getting greater value for money. Wherever possible, partners should be appointed for a series of projects. Then, as the arrangement develops, clear improvement targets should be set. Open-book accounting should be used so departments have an assurance about contractors' costs and efficiency improvements.

A crucial decision that influences this stage is the way the price will be established. The main options used in the construction industry are:

- Cost reimbursement, which is entirely consistent with partnering because it allows firms to concentrate on delivering best value within the constraints of the client's objectives including any overall budget. It is important that the procedures and systems used to calculate the price are open, robust and subject to audit.
- Schedule of rates, which is consistent with partnering because it provides similar flexibility to cost reimbursement for work covered by the schedule of rates. The price for work outside the schedule of rates has to be negotiated and most standard forms of contract rely on a cost reimbursement basis. So where the firm is to be fully involved in searching for best value it often makes sense to avoid the costs involved in using a schedule of rates and use cost reimbursement.
- Bills of quantities can be used with partnering but lack the easy flexibility of cost reimbursement and schedule of rates. Major changes to the design that improve value for money are not always priced fairly by bills of quantities especially where the bills do not include relevant unit prices. The prices have to be negotiated and most standard forms of contract rely on a cost reimbursement basis. So where the firm is to be fully involved in searching for best value it makes sense to use cost reimbursement.
- Lump sum is inconsistent with partnering. It can be used on partnering projects for minor elements or systems that are fully defined where the firm providing them will not be brought into the partnering arrangement.

Many clients wanting to use partnering will ensure that all consultants, contractors and specialists invited to submit tenders also want to use partnering. Some clients want to explore all the possibilities and decide that the tender documents should include a clause that invites tenderers to state whether they wish to enter into a partnering arrangement with the client if they are selected. This recognizes that partnering must be voluntary. Having invited a firm to submit a tender, a decision that they do not wish to partner should have no influence on their chances of winning the contract beyond the agreed evaluation criteria. This means that such a firm can provide convincing evidence that they will provide the client with better value for money than that offered by any firm willing to use partnering. This is unlikely but it can happen.

It is often sensible to conduct pre-tender briefing meetings at which the formal tender documents are issued. Key principles in ensuring there is fair and transparent competition in a single round of tendering consisting of either one stage or two stages include:

- All tenderers are provided with exactly the same information which is sufficient to prepare directly comparable tenders.
- All parties respect each other's confidentiality.
- Tenderers are allowed sufficient time to prepare properly considered tenders.
- There is no collusion between tenderers.
- Tenders are evaluated on performance and price.
- Tenderers are not pressured into reducing their price.
- Alternative proposals that may provide better value for the client can be offered in addition to a tender that fully complies with the issued tender information.
- All tenderers are informed of the outcome promptly.

The firms invited to submit tenders should be given information which includes:

- Instructions to tenderers telling them how, where and when their tender must be submitted.
- The type and form of tender (priced bills of quantities, a schedule of rates, a basis for cost reimbursement, or some combination of these), together with other information, particularly the names of the client and any project team members already appointed.
- Whether they are allowed to submit alternative proposals in addition to a tender that complies with the instructions.
- How the tenders will be evaluated including particularly the weight given to performance and price.
- A description of the required building, either as a minimal statement or outline design, and arrangements for visiting the site during the tender period.
- The terms of the contract that will be used.
- The date the site will be available to the contractor and the required completion date.
- The payment terms describing interim payments and retentions.
- Insurance provisions provided by the client and required from the tenderer.

Best practice balances performance against price as described in checklist 6.

8 Decision-Making Systems

Decision-making systems have a number of distinct elements described in checklists 9 to 16. Together they provide an overall decision-making system for construction projects which is illustrated in Figure 7.1.

The separate elements of decision-making are normally taken for granted inside individual firms. However, partnering requires them to be coordinated so that they support the project team as they cooperate in making decisions. Each element should be considered at the first partnering workshop with this aim in mind.

In designing their decision-making system, project teams should look for ways of streamlining the administration of their project. So for each element of the decision-making system they should ask:

- Is this activity absolutely necessary?
- Does it duplicate what others are doing?
- Can it be merged with some other activity?
- How can this activity be simplified, made more reliable or quicker?
- How can this activity contribute to making essential activities more certain, efficient, faster, safer or better quality?

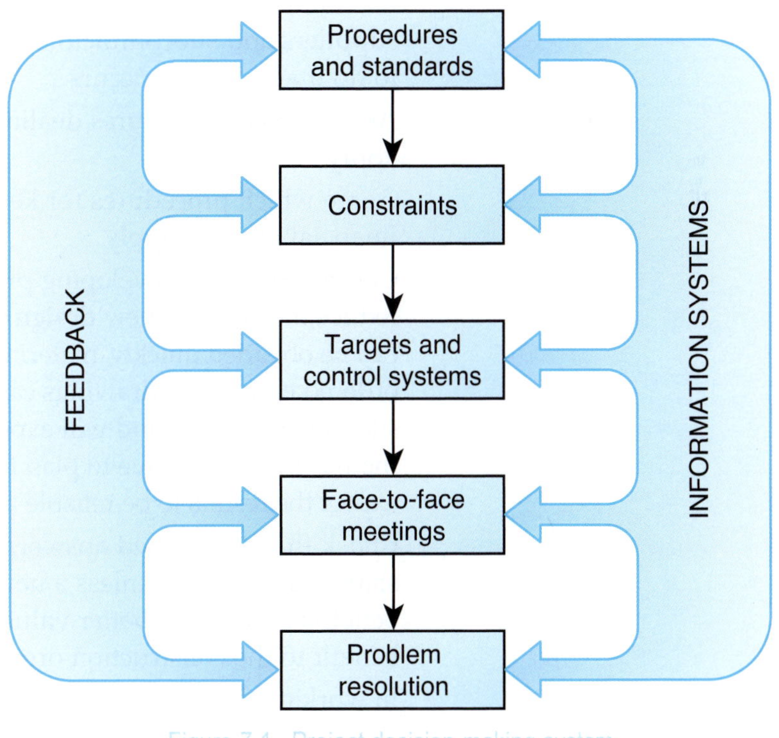

Figure 7.1 Project decision-making system

9 Procedures and Standards in Partnering Projects

Project teams should:

- Review the procedures and standards they will use to ensure they make a positive contribution to the efficiency or effectiveness of project work.
- Agree which public procedures and standards apply, e.g. standard forms of contract, methods of measurement and coordinated project information.
- Review any internal procedures and standards used by any of the work teams that may influence other work teams, e.g. a standard approach to design or how various categories of information should be prepared, presented and checked.
- Check the compatibility of computer-based systems including the way they handle various types of information.
- Agree which standard design details apply and check that any related information about price, construction implications, maintenance information and environmental impact is relevant.
- Agree which procedures for production processes apply including the arrangements for dealing with joints, junctions and fixings between the work of separate work teams.
- Agree which procedures for the behaviour of staff apply including workers' behaviour towards each other, customers, suppliers and subcontractors, how to act in a crisis and what to do if an accident occurs.
- Agree which procedures dealing with matters of discipline apply.
- Agree which procedures for keeping records, particularly financial records, apply.
- Consider ways of developing procedures or standards that reduce the need for new design work so that official approvals can be obtained quickly, materials and components can be ordered early, shop drawings can be reused, construction on site can be efficient and will experience few problems, and there is every incentive to plan for fast, accurate work and to expect the results to be reliable and efficient.
- Check that established answers embodied in procedures and standards are used unless a new answer provides significant benefits in terms of better value to the customer or a specific benefit to the construction organization.
- In working through the above items, use the box opposite as a checklist of issues that may be covered by procedures and standards.

Basic Checklist of Issues that can be dealt with by Procedures and Standards

- Planning regulations
- Technical regulations
- Nuisance regulations
- Working conditions
- Designs
- Specifications
- Contracts
- Employment conditions
- Discipline
- Project information
- Construction methods
- Health and safety
- Accidents
- Quality records
- Cost records
- Financial accounts
- Payments
- Programmes
- Performance measures
- Communications
- Feedback channels
- CDM regulations
- Sustainability.

10 Constraints in Partnering Projects

Project teams should:

- Review the constraints that define the levels of performance they have to achieve.
- Agree which legislation applies, e.g. health and safety legislation.
- Agree which official regulations apply including any that restrict forms of construction, methods that can be used, safety requirements and quality standards.
- Agree which constraints arise as an integral part of procedures and standards adopted by the project team or any of the work teams.
- Agree work teams' interpretation of regulations, procedures, standards and norms, e.g. is zero defects a target to aim for or a constraint that will be achieved?
- Agree work team's approach to planning and controlling project work, e.g. is it normal to produce detailed construction plans and put them into effect without any changes or to treat planning as a flexible tool that has to respond to events and changes as projects proceed?
- Agree how far well-established supply chains can be used so that the sequence and timing of every activity, design details and methods of construction can be selected with confidence.
- In working through the above items, use the following box as a checklist of issues that may be covered by constraints.

Basic Checklist of Issues that may be Treated as Constraints

- Safety
- Quality
- Efficiency
- Programme
- Budget
- Open communication
- Trust.

11 Targets and Control Systems in Partnering Projects

- Agree the targets that apply to the project team.
- Agree the targets that apply to individual work teams.
- Work teams in the early stages of partnering relationships need targets that can be measured objectively so that there is no basis for arguments about the accuracy or relevance of feedback.
- All targets need to be compatible with agreed mutual objectives and performance improvements.
- Work teams should concentrate their efforts on a small number of simple, measurable targets that are of central importance to the project's success.
- The project team needs reliable information about the best performance being achieved by other teams tackling similar projects in similar circumstances.
- Constructing Excellence in the Built Environment's key performance indicators (see box below) provide a good basis for ensuring that targets are set at appropriate levels.

The general relationship between targets and control systems in construction work is illustrated in Figure 7.2.

- Control systems need to be in place to provide project teams and each work team with objective information about their own performance related to their targets.
- Control systems should give an early warning when work deviates from an agreed target that is sufficiently specific to tell the people involved where to concentrate their efforts.

Constructing Excellence in the Built Environment's Key Performance Indicators

- Client's satisfaction with the product
- Client's satisfaction with the services
- Defects
- Costs
- Predictability of costs
- Time
- Predictability of time
- Safety
- Productivity
- Profitability.

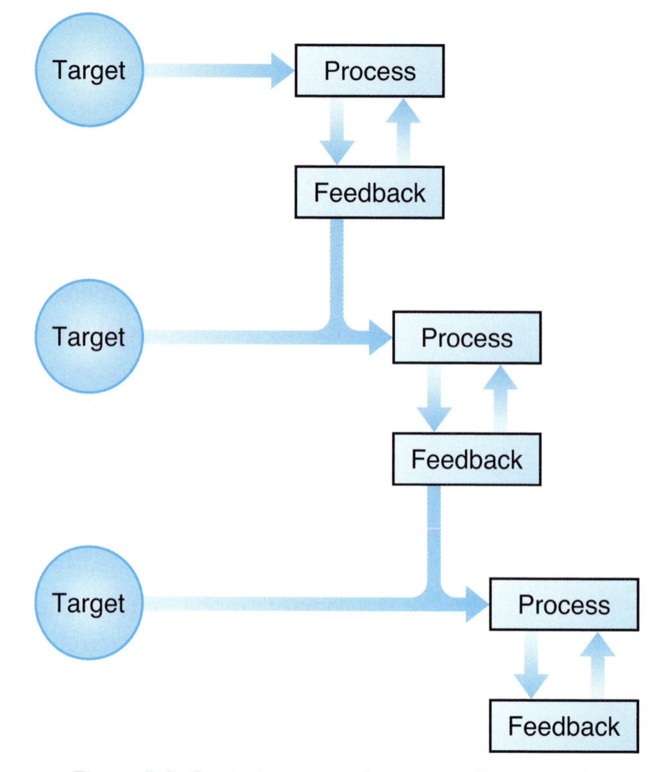

Figure 7.2 Control systems in construction projects

- Feedback needs to be a living thing, subject to constant discussion by work teams rather than merely the subject of infrequent formal reviews by managers outside the team.
- Feedback needs to be visible, shared with everyone and supported by training, so that everyone understands how it is produced and what it means for the success of the project.
- Check that work teams understand and agree the measurements of their own performance so feedback can be discussed by the team.
- Check that where no well-established measures already exist, teams are involved in developing measures of their own performance.
- Check whether established benchmarks can be used to make meaningful comparisons of the performance of individual projects. These may be based on any of the measures listed in the box opposite.
- Feedback must be accurate and up to date.
- Check that teams have all the support they need to give them every chance of achieving their targets.
- All teams should act to solve problems and not look for excuses for failure.
- Check that when a key target is in danger of being missed, this is treated as a crisis and clear effective action is taken quickly to get the work back on its planned course.
- Senior managers need strategic feedback including progress, productivity, profit levels and environmental and social impacts.

Measures that may Provide Benchmarks of Project Performance

- Energy use of new buildings
- Cleaning and maintenance costs of new buildings
- Number of complaints from clients or neighbours
- Number of real innovations introduced
- Percentage number of targets achieved
- Number of revised design drawings issued
- Number of change orders issued
- Proportion of rejected materials or components delivered to site
- Number of defects on handover
- Time taken and the costs involved in dealing with defects
- Cost of a given unit of completed work
- Deviations from cost targets
- Time taken to complete a standard unit of construction
- Deviations from time targets
- Speed of getting back on programme after a delay
- Number of injuries per 1,000 man/days worked on site
- Quality of welfare and safety provisions
- Percentage number of hours devoted to training
- Percentage of time spent by teams each day on key objectives
- Proportion of people at meetings directly involved in the subjects being discussed
- Time taken to respond to requests for information from other project team members.

12 Meetings for Partnering Projects for Standardized Solutions

This technical description suggests the meetings needed to provide an effective framework for projects using standardized solutions that are well understood by the project team, which ideally is based on site. Some of the meetings can be combined or held informally on small or straightforward projects.

Work team meetings

All work teams meet as often as they need to complete their work.

In addition the following meetings involve more than one work team.

Milestone meetings

The core team meets as each milestone is reached (typically one a month) to ensure that the current milestone has been met, to review performance and plan the work leading to the next milestone.

Supply chain meetings

Each supply chain team meets with the key members of the direct work teams currently involved in the work every week to review progress, to look for better ways of working and resolve any problems. These meetings take place as long as work on the particular element or system is still under way.

Construction meetings

Key members of the core team and site supervisors currently working on site meet daily to review the day's progress; solve any problems affecting this progress; and plan the next day's work in detail. These meetings are held right through the construction stage.

13 Meetings for Partnering Projects for Original Designs

This technical description suggests the meetings needed to provide an effective framework for projects producing original designs. It assumes that the project team works in a number of different places including design offices, factories and on site. Some of the meetings can be combined or held informally on small or straightforward projects.

Work team meetings

All work teams meet as often as they need to complete their work.

In addition the following meetings involve more than one work team.

Milestone meetings

The core team meets as each milestone is reached (typically one a month) to ensure that the current milestone has been met, to review performance and plan the work leading to the next milestone.

Sub-stage meetings

The core team meets every week to solve any problems not being dealt with elsewhere. They review the design, construction methods, progress and cost reports and external influences to ensure that the project is meeting agreed targets and make decisions about problems and opportunities.

Design meetings

The core team and key members of the supply chain teams currently involved in design meet every week. They review the design, any feedback from the client's internal team and supply chains, explore alternative solutions, ensure that the design is meeting all the agreed targets, and resolve any design problems. Once all the design decisions are made, these meetings cease.

Supply chain meetings

Within each supply chain representatives of all the work teams currently involved in the project meet every week. They ensure that all necessary information is in place for the coordination of the design and for efficient construction, look for better ways of working and resolve any problems. These meetings take place as long as work on the supply chain is still under way. Members of the core team attend key meetings.

Direct work meetings

Key members of the core team meet every week with the contract manager and site supervisor of the lead firm in each supply chain. Initially each such meeting resolves any design and procurement problems, agrees the exact scope of the work, agrees the terms on which it will be carried out, including ensuring there is a fair basis for payment, and checks that the work meets all the agreed targets. Once these decisions are made, subsequent meetings concentrate on ensuring that detail design, construction method statements and construction are meeting agreed targets, and resolve any problems. The meetings are held as long as the lead firm is working on the project.

Construction meetings

Key members of the core team and site supervisors currently working on site meet every day. They review the day's progress; solve any problems affecting this progress; and plan the next day's work in detail. These meetings are held right through the construction stage.

- Check that the databases and formal flows of information available to the project team identify relevant information and distribute it to everyone who needs to know, in a convenient form, at the appropriate time.
- Check that people are not being bombarded with copies of information they do not need. If this is happening, the core team must act to stop this sloppy way of working.
- The client needs to have good information about the value being delivered and its progress towards achieving agreed objectives.
- All teams need to have good information about the client's objectives, the current design, construction plans as well as their own quality, time, cost and safety performance.
- All teams need to know what they are required to do in sufficient time to plan and organize their work.
- Teams need to know how their work relates to that of other teams.
- Check that information systems make use of modern computer systems, e.g. that computer-aided design systems are linked to expert systems that provide construction management information.
- Information should be organized so that it can be accessed in text, graphics, sound, video or whatever form the users find most convenient.
- Check that the categories of information listed in the box below are dealt with by the project's information systems.

Categories of Information Needed by Project Teams

- Products and services
- Processes
- Constraints
- Targets including how they are measured
- Control systems including feedback
- Problem-resolution support.

15 Feedback Systems for Partnering Projects

- Check there is systematic feedback available to guide the performance of every team involved in the project.
- Check whether the feedback systems need to be developed.
- Check whether teams need training in the use of feedback systems.

Useful Measures of Project Team Performance

- Percentage of work samples meeting specified quality standards
- Percentage of work items within cost target
- Percentage of activities on programme
- Number of reportable accidents
- Percentage of meetings starting on time with everyone present
- Number of days training provided
- Number of defects at handover
- Percentage of users satisfied with the new building or infrastructure
- Building or infrastructure cost as percentage of industry norm
- Project time as percentage of industry norm.

- Senior managers should be involved in deciding the balance between cost, time and quality in the feedback systems used by work teams and project teams.
- Work teams should be directly involved in establishing their own targets taking account of the overall mutual objectives and performance improvements agreed by the project team.
- Check that the feedback systems use data collected by work teams about their own performance that they understand and accept as a fair measure of their performance.
- Work teams should use feedback to control their own work by comparing their performance with the targets.
- Check that the feedback systems give the core team information describing the medium-term trends in performance, highlighting risks and uncertainties and providing details of major sources of interference with planned progress or methods.

- Check that senior managers and members of core teams visit the offices, factories and construction sites where project work is taking place and ask questions to help interpret information provided by feedback systems.

Questions Senior Managers can Ask During Visits to Workplaces

- What are your current targets?
- How well are you doing against your targets?
- How well did you do last week?
- What is the next project milestone you have to meet?
- What quality standards are you working to?
- Which other work teams depend on your outputs?
- How do you keep them informed about progress?
- Do you know who to ask for any information you need?
- What performance improvements are you working to achieve?

- Members of core teams should work with an open door, in the sense that they will talk to anyone who comes to their office with a problem.
- Consider the use of a suggestions box particularly on large projects where anyone can put a note, a copy of a document, a comment, a suggestion or anything they think the core team should see.
- Ensure that a suggestions box is not used as an excuse to remove the responsibility of work teams for dealing with problems quickly.
- Check that the project team and work teams at all levels are relentlessly determined to achieve all their targets and search for performance improvements.

16 Procedures for Resolving Problems

Overall aims in resolving problems

- Find permanent solutions quickly.
- Maintain normal planned and controlled work unless this is impossible, in which case ensure it is resumed as soon as possible.

Decision-making framework

- Agree a carefully defined question and identify the criteria for satisfactory answers.
- Check that the question relates to actual physical activities and outcomes, not what is going on inside people's heads.
- Check that the question treats the problem as an opportunity for finding better ways of working.
- Use creative techniques to identify a number of possible answers.
- Consider the possibility that good answers already exist elsewhere or at least ask if other teams have faced similar situations.
- Check that the project's information systems allow the crisis or problem situation to be described so that people with directly relevant answers or experience can be identified.
- Evaluate the two or three most promising answers by judging them against the agreed criteria, and listing the strengths and weaknesses of each, taking account of the consequences on other elements of the work.
- Check that in judging answers against the criteria teams use fair procedures.
- Select one of the evaluated answers; adopt a combination of elements from the potential answers; evaluate more of the potential answers; search for more answers or review the original question.
- The overall structure of the procedures for resolving problems should be based on that used in best practice decision-making shown in Figure 7.3.

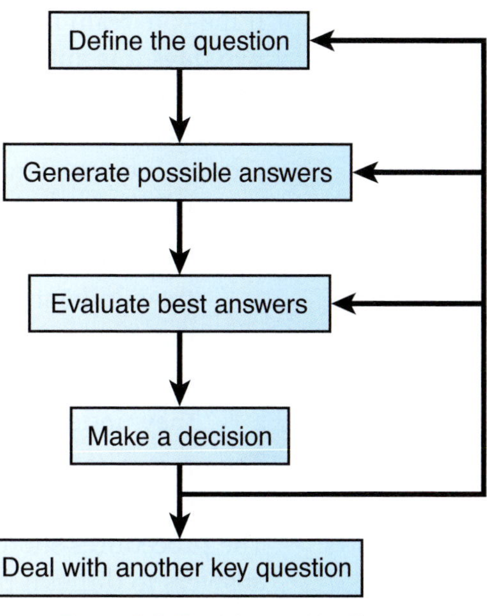

Figure 7.3 Decision-making framework

Problem-solving actions and attitudes

- Teams look at their own contributions to problems and do not try to blame others.
- Teams concentrate on what they can do to help solve the problem. They do not try to tell others what they should do because that implies blame and creates conflicts.
- Teams look at problems with a wide perspective and different points of view.
- Teams look for the positive aspects of any situation and in looking for answers concentrate on reinforcing things that work well.
- Teams recognize that if some situation really is bad and they can do something about it or learn some lessons from it, they should do so and then move on to the next task.
- Teams recognize that if there really is nothing that can be done about a bad situation, they just accept the inevitable and do not waste time and energy dwelling on it.

Problem-solving procedures

- Problems should be resolved by the people directly involved who understand the issues and have all the data needed to find a good answer.
- Set tight time limits for the people directly involved to find an answer that enables them to keep their work within its defined constraints and targets.
- If the people directly involved are not able to find an answer within a predetermined and short time limit (two days is sensible) the problem is referred to the core team.

- If the core team is unable to find an answer within a further two days it is automatically referred to a group of senior managers from the partnering firms to be resolved quickly (three days is sensible). The criteria used in appointing the group of senior managers should include:
  - they are available at short notice
  - they are committed to the project meeting its targets
  - they understand partnering
  - they are competent to represent the interests of the firms affected by their decision
  - they understand the problem or crisis.
- The procedures used in resolving problems should broadly match the pattern shown in Figure 7.4.
- The procedures should include the use of task forces to find answers to difficult problems.
- On major projects the procedures should allow two or three task forces to work simultaneously on the same problem and

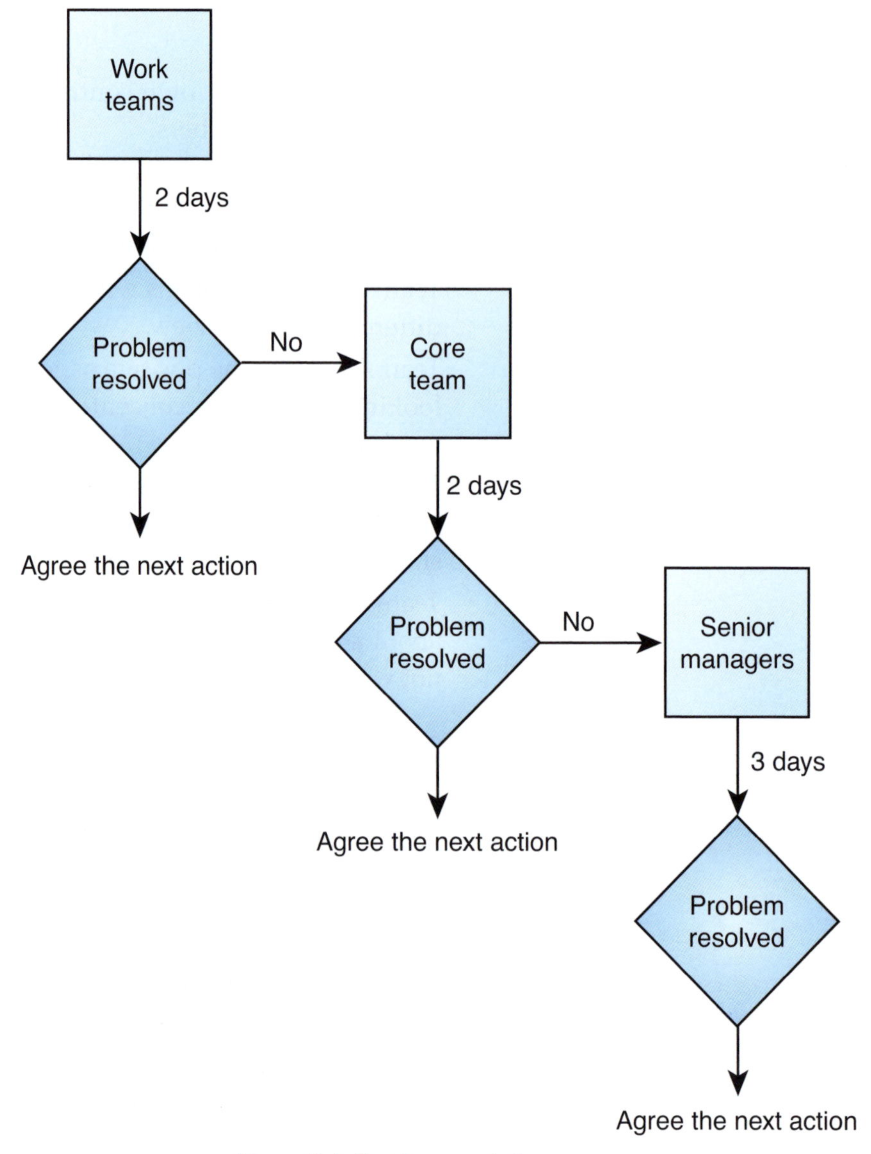

Figure 7.4 Problem resolution process

present their answers in front of each other so they can be discussed in the search for a good answer.

- The procedures should allow for a partnering workshop to be used to deal with persistent and difficult problems so that different interests can be taken into account in searching for an answer that ideally gives everyone all they really need, and as a minimum can be regarded as fair and reasonable.
- The procedures should provide strong counter measures to deal with individuals that act selfishly against the interests of the rest of the project.
- Counter measures should be clear and certain so that no one believes that they can exploit the project team for their own advantage and get away with it.
- Counter measures need to be commensurate with selfish acts and may include detailed checking and additional audits of the person's work, a requirement for frequent progress reports, or other similar routines that focus on the interests of the whole project.
- Counter measures should be strengthened and made more urgent if selfish acts are repeated.
- If the selfish acts continue in the face of the strengthened counter measures, the procedures should provide for the individual to be replaced on the basis that everyone deserves a second chance but not an endless stream of chances.

17 Performance Improvements

- Firms involved in partnering need to invest in the continuous improvement of their actions, products, processes and systems.
- Project teams and work teams need to be committed to cooperative teamwork.
- Work teams need to respect differences, build on strengths and compensate for weaknesses.
- Work teams should encourage people to describe their real needs and feelings, work at understanding them and look for answers that give everyone more of what they need.
- Work teams should encourage people to be open about new ideas, feelings and experiences.
- Work teams need to value different points of view and try to understand them because they may provide a basis for a new answer that improves performance.
- Work teams need to be competent in process analysis, work measurement and basic statistical techniques or training needs to be provided.
- Work teams should regularly consider what they can improve by identifying problems that need to be tackled.
- Work teams deciding on improvements should take account of the impact of changes on clients.
- Work teams need to analyse the processes involved in the problems they tackle by identifying the stages, all the inputs and outputs and the causes and effects that determine performance. Figure 7.5 shows a typical pattern for the resulting process analysis diagram.

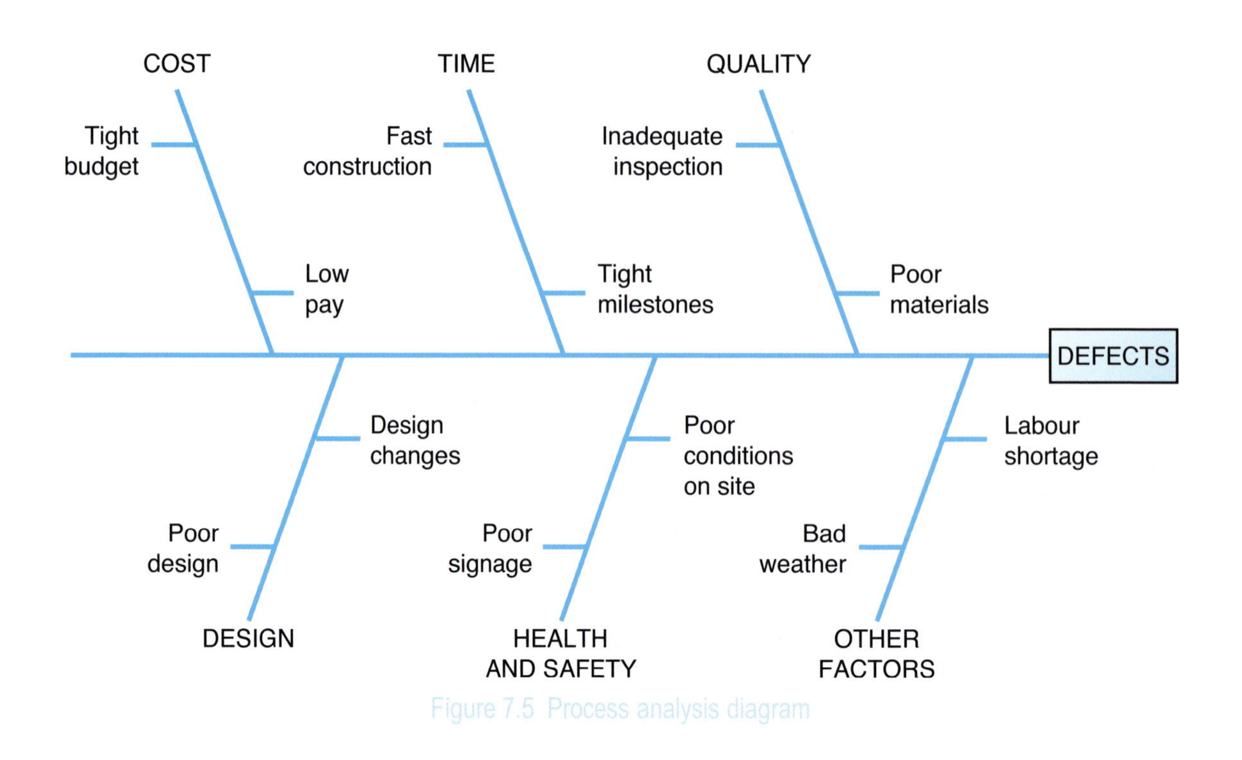

Figure 7.5 Process analysis diagram

- Work teams aiming to solve a problem need to measure the key features of the processes involved to establish the extent and location of problems and opportunities for improvement.
- Work teams should set a measurable target for improving the existing situation.
- Work teams need to agree specific actions designed to achieve the planned improvement and put the actions into practice.
- Work teams should monitor the effects of the changes by means of further measurements of their performance.
- If the actions do not solve the original problem, the work team needs to review the earlier steps to look for more effective actions.
- Work teams should measure their own improvements in performance and regularly report their achievements to senior managers so that everyone knows whether they are doing better this year than last and by how much.
- Senior managers need to understand that incremental improvements cannot continue forever and that all technologies have a life cycle which takes the general form of a sigmoid curve illustrated in Figure 7.6.
- Senior managers need to understand that the life cycle of most technologies begins slowly, experimentally and falteringly, moves through a period of rapid growth and expansion to a final slowing down and decline.
- Senior managers should know where the technologies used by their organizations are on the life-cycle curve so they can start investing in new technologies during the middle phase of rapid growth, well before the decline sets in.

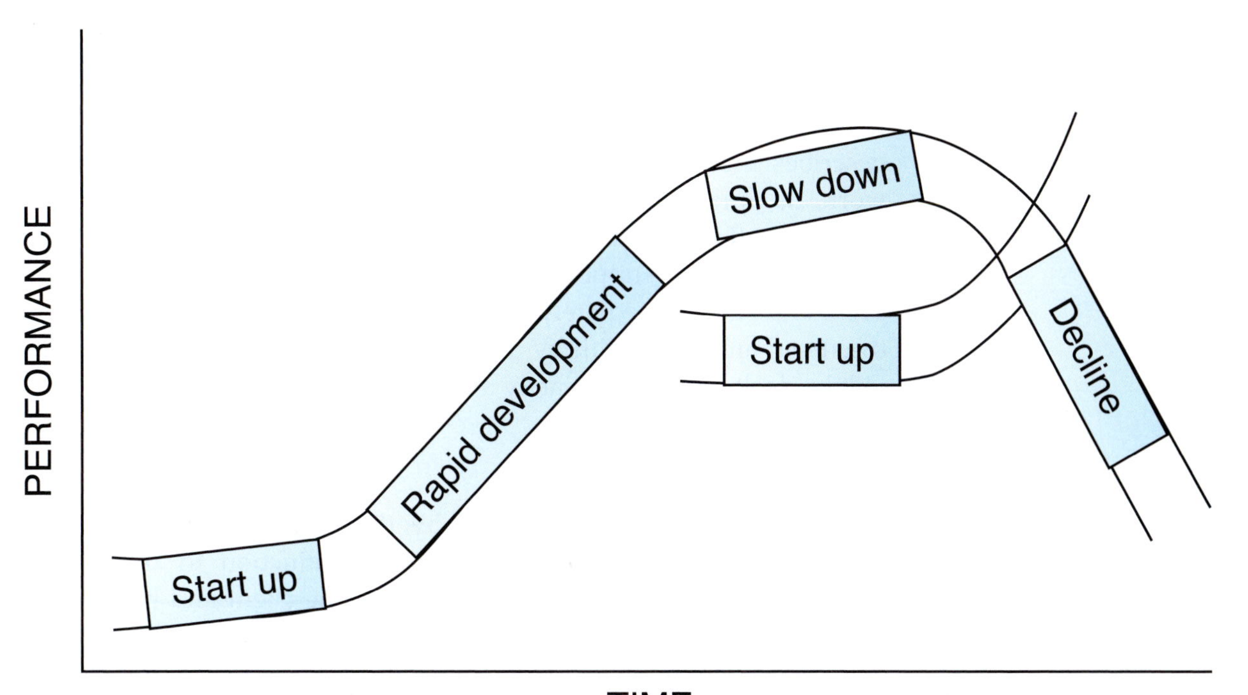

Figure 7.6 Life cycle of technologies

18 Induction Courses

- Induction courses are held for each group of work teams before they come into a common project office or onto site.
- Induction courses bring together the group of work teams that will be working together and includes key members of the project core team.
- Induction courses tell people who will be working together about the project, its design and the way work is organized, reinforcing the project's use of partnering and cooperative teamwork.
- Induction courses normally take one day and it is usually preferable for the course to be run by an independent facilitator.
- Induction courses are best held in the common project office or on site but when no suitable space exists they should be held in a neutral location that allows plenary sessions and small group discussions.
- A preparation meeting should be held two weeks before the induction course takes place, attended by the facilitator, the members of the core team and the leader of each work team that will attend the induction course.
- The preparation meeting agrees the form of the induction course and agrees who will deal with each of the following subjects.
- During the induction course everyone should be encouraged to raise concerns or questions and these should be dealt with thoroughly and openly.

The general form of induction courses is as follows.

Introductions

Everyone attending the induction course introduces themselves and describes their role. Each work team leader describes their team's work, its timing, the composition of their work team, the plant and equipment they will use, the main materials and components that will be delivered, the support services they need and any outstanding problems. Any immediately obvious clashes or interface problems are discussed and dealt with by making a decision or agreeing how they will be dealt with immediately after the induction course.

The project

The new facility being produced by the project is described. This should provide an exciting inspiring introduction to the project, emphasizing the way it delivers value for the client and how it contributes to the viability and competence of the consultants, contractors and specialists involved. This should be a very visual presentation describing the design, including the overall design philosophy and its impact on the local environment. The presentation explains how the new facility benefits the client's business. The presentation should emphasize any direct implications of the design for the work teams at the induction course. These may include quality, workmanship, tolerances, care of completed work and other issues.

Project team

The project team and the way it has agreed to work should be described. The presentation should describe the members of the project team that the work teams at the induction course are likely to come into contact with.

Project partnering

The way project partnering is being put into effect is described. This should describe the project's mutual objectives, decision-making systems and performance improvements. If there is an agreed partnering charter, this should be issued to everyone present and discussed. Any partnering workshops that the work teams at the induction course will be involved in are described. The presentation should emphasize the use of cooperative teamwork and describe the arrangements set up to ensure that problems are solved quickly. It should emphasize the aim of achieving specific performance improvements. It should also emphasize that the financial position of every firm involved is secure insofar as this is directly influenced by the project. The presentation should emphasize the benefits of partnering and that the project provides an important opportunity for everyone involved to develop their own skills and knowledge in ways likely to benefit them on future projects.

Project progress

The stage the project has reached is described. This, like all the presentations, must be an entirely honest description. Ideally this is an inspiring presentation describing the excellent quality, progress, efficiency and safety being achieved. It should emphasize the performance improvements being achieved. In different circumstances, the presentation will explain how and why the project is failing and the actions being taken to ensure that the project will be fully complete within budget by the agreed completion date.

Work places

The work places where the work teams will carry out their tasks are described. This presentation describes what work has already been done in those locations so the work teams know what conditions will exist when they begin their own work. It describes how completed work is to be protected. It describes how materials are delivered, stored and distributed. It describes the use of common plant and equipment. It describes the health and safety provisions that apply. It describes which work teams in addition to those at the induction course have access to the work places. It describes how the work places will be managed while the work teams are carrying out their work.

Information

The way information describing the project work is produced, distributed and updated is described. The presentation should emphasize that everyone is responsible for ensuring that they are working on up-to-date information and for raising any problems immediately they become aware of them.

Health and safety

The health and safety coordinator and principle contractor should describe the main features of the health and safety plan, emphasizing the provisions that directly affect the work teams at the induction course. The presentation should emphasize the benefits of safe work and the sanctions that apply to workers breaking health and safety rules. It should particularly emphasize the benefits of keeping the site clean and tidy and removing all rubbish every day. The presentation should describe the cafeteria, rest areas, first aid, toilet and other welfare provisions.

Quality control

The quality control systems are described. The presentation should highlight any particular quality requirements that affect the work teams attending the induction course. It should include detailed information about the type and frequency of inspections and tests. It should emphasize the benefits of reliable good quality to everyone involved and the waste involved in redoing work and remedying defects. 'Right first time' is a useful slogan and attitude on construction projects. The presentation should explain the benefits of good quality to the client.

Time control

The way information describing the timing of project work is produced, distributed and updated is described. The presentation should show the work teams the work programmes they will use and describe the way progress is monitored. It should particularly describe information the work teams are required to produce and explain how this contributes to the time control systems being used. The presentation should emphasize the benefits of sticking to programmes and describe the problems caused to other work teams when work is not fully completed on time. It should also describe the importance to the client of finishing the project exactly on time.

Cost control

The way information describing the cost of project work is produced, distributed and updated is described. Also the way cost is monitored and the cost control systems being used are described. The presentation should emphasize the benefits of costs being controlled for the client, consultants, contractors and specialists including the impact on profits.

Project meetings

The pattern and timing of project meetings that involve work teams attending the induction course are described. This includes the way meetings are run, emphasizing the importance of people attending having done their homework and having the authority to make decisions. The way work teams should report problems and concerns is described. The presentation should emphasize the role of the daily construction meetings that bring together the core team and site supervisors currently working on site. The presentation should also emphasize the importance of work teams sorting out problems themselves quickly and directly. Subjects and issues on which work teams are expected to make their own decisions and those which must be referred to the core team are described.

Team building

A number of practical exercises should be undertaken in small groups. These should require participants to discuss a practical issue directly relevant to their work. Ideally the results will identify a better way of working that can be used on the project. Whatever the results, they should be reported back to a plenary session. When all the groups have reported back, the results should be discussed positively with the explicit aim of encouraging cooperative teamwork by demonstrating that good ideas may come from any member of the team.

Conclusions

The outcomes of the induction course are listed, discussed and agreed. This part of the induction course should explicitly consider the need for training or development identified during the induction course. Specific actions should be agreed to deal with individual or team needs. All follow-up actions are agreed including deciding who will take each action.

19 Partnering Workshops

- The specific arrangements for each partnering workshop depend on the stage the project has reached and are worked out by the core team using the following general guidance.

Partnering facilitator

- A partnering facilitator should be appointed in time for essential preparation to be carried out thoroughly. The partnering facilitator usually comes from outside the project team although experienced partnering teams can use an internal facilitator.
- The skills needed by partnering facilitators are similar to those used by facilitators in workshops that form part of formal negotiations, value management, risk management, team building, benchmarking, etc. They need to understand the principles of partnering, and it helps if they have knowledge of construction.
- The partnering facilitator will be assisted by other facilitators, the number depends on the number of people attending (typically one facilitator for every six to eight people).

Workshop objectives

- The core team agrees with the facilitator the workshop objectives in general terms. These are made more specific during the preparations for the workshop. The objectives depend on the stage a project has reached as described in Chapters 3 and 4.

People involved

- The list of those to attend should be agreed by the core team with the facilitator to include everyone likely to make a significant contribution to the workshop's success. They should all have the authority to make decisions that will be supported and put into effect by their own firms. The following should be considered:
  - Project sponsor
  - Client's facilities manger
  - Client's finance director
  - Lead designer
  - Lead construction manager
  - Managers responsible for quality, time and cost control systems
  - Health and safety coordinator
  - Managers responsible for each technology cluster
  - Representatives of all the stakeholders.

Timing

- Partnering workshops normally take two days and are held at a venue that ensures that those attending are not interrupted. People attending should be given as much notice as possible so they can keep the time free of other commitments.

Keeping a record

- A record should be kept of every good idea, all the decisions and any hints of problems. This is often best achieved by all the facilitators making notes and one of them having a specific responsibility for producing the record. This responsibility can rotate amongst the facilitators.
- There may be circumstances, particularly on large complex projects, where more extensive records are needed. This should be considered and the precise methods to be used agreed. These may include audio and video recordings, overhead projector slides and material presented using computer technologies.

Preparation

- Preparation begins well before the workshop because on most projects there are too many issues to leave them all for the workshop. Many issues can be decided beforehand and reported to the workshop, which gives an opportunity for decisions to be questioned and, if new information makes it necessary, for them to be changed. Where the matter needs discussion at the workshop, the background information should be assembled and distributed beforehand. The aim is for the workshop to concentrate on issues where no satisfactory answer is obvious so it needs to be discussed seriously and an agreement found.
- The partnering facilitator interviews all the people attending the workshop either face-to-face or by telephone to ensure they understand the purpose of the workshop and how it will be run. The facilitator should ensure that people attending have the authority to make decisions that commit their firms. The facilitator should identify problems and concerns so they are discussed at the workshop. The facilitator should check that the workshop includes a variety of personalities and views so that assumptions will be challenged and new ideas introduced.
- The partnering facilitator should check that all the subjects to be dealt with at the workshop are considered beforehand and identify what is already agreed and what remains to be decided. In doing this, good facilitators ensure that the workshop is not being railroaded by powerful individuals into ill-considered agreements.
- The facilitator should identify during the interviews whether any of the participants would benefit from coaching or

training in making presentations or other aspects of the workshop. This should be discussed with the core team and suitable arrangements made.

- The facilitator should prepare a flexible agenda for the workshop to provide the basis for a handbook issued to participants. The handbook will draw on libraries of techniques and games taking account of the specific project, the participants and the objectives. The handbook can usefully include background reading that helps in understanding the workshop objectives and the approach being adopted to achieve them.

The venue

- The facilitator should visit the workshop's venue well before the event and undertake the following actions.
  - Check that it provides for plenary sessions and small group discussions.
  - Agree the furniture type and layout, check that the heating and lighting levels are comfortable and make sure there will be no interruptions from noise or other distractions.
  - Check the right visual aids will be provided and will work effectively. These should include flip charts, overhead projectors and modern electronic technology. It should be possible to display flipchart sheets around the workshop room.
  - Check that the ideas for recording the workshop discussions and decisions will work reliably.
  - Check that participants can walk outside in fresh air and attractive surroundings.
  - Check that water, tea and coffee are available throughout the two days and that the arrangements for breakfast, lunch and dinner will support the workshop's progress.
  - Check that everyone can stay overnight at the venue so that work can continue reasonably late, the participants can have a good dinner together and continue talking as late as they wish.
  - Check the arrangements to ensure an early start on the second morning, which is often the most productive time. This is because when people become absorbed in a subject and have an opportunity to sleep on it, many of them get their best ideas the next morning.
  - Deal with any problems by insisting that if any of the provisions are not satisfactory, they must be improved so that the venue gives the workshop every chance of being enjoyable, productive and successful.

Formal invitation

- Send a formal invitation to all the participants about a week beforehand that includes:
  - A statement of the client's objectives for the project
  - The venue and times
  - The agenda and objectives for the workshop
  - The procedures that will be used
  - The name and role of the partnering facilitator
  - A point of contact for raising queries.

The workshop

- The workshop is guided by the facilitator using a variety of decision-making techniques and games flexibly to achieve the agreed objectives and foster cooperative teamwork.

Typical Outline Agenda for a First Partnering Workshop

Day 1

1. Welcome and introductions
2. Discussion of how individual styles and personality traits affect working relationships
3. Presentation on cooperative teamwork
4. Game to demonstrate the benefits of cooperation
5. Team-building exercise based on discussing what each participant needs from the project
6. Game based on identifying team strengths
7. Discuss mutual objectives
8. Brainstorm major problems or obstacles and discuss answers
9. Discuss performance improvements
10. Agree tentative conclusions on mutual objectives and performance improvements.

Day 2

1. Review mutual objectives and performance improvements
2. Presentation on cooperative teamwork and dealing with problems
3. Review major problems or obstacles
4. Game to devise the problem-resolution process
5. Discuss decision-making

Continued

Typical Outline Agenda for a First Partnering Workshop – cont'd

Day 2

6. Game to agree the decision-making process
7. Agree the decision-making and problem-resolution process
8. Discuss and agree the Partnering Charter
9. Close.

- Partnering workshops depend on the skillful use of various games by the facilitator as the workshop progresses. The following are typical games.

Getting to know you

This provides an alternative to having each person introduce themselves. The facilitator arranges the participants into pairs and gives them ten minutes to interview each other. Then each person introduces the other person to the workshop stating:

- Who they are
- Who they work for
- Their role in the project
- Their experience of partnering
- Their main interests
- An interesting fact about them.

Your real interests

This game can be used to identify people's real interests. Pairs of partners discuss a selection of the following or similar questions and report their partner's answers to the workshop.

- What is the best thing about this project?
- What worries you most about this project?
- What does this project have to provide for your firm to regard it as a success?
- What do you hope the project provides?
- What do you expect the project to provide?

Cooperation works

An important example of good partnering games is provided by the Carlisle Partnership's red/blue exercise. It creates an opportunity for workshop participants to experience the implications of cooperating or competing. It is based on the prisoners' dilemma.

The prisoner's dilemma requires two parties to make independent decisions about whether to cooperate or compete. There are four outcomes, for which the table below shows a typical set of payoffs which create a dilemma for participants.

		Decision by A	
		Cooperate	**Compete**
Decision by B	**Cooperate**	Both get 3	A gets 5 B gets 0
	Compete	A gets 0 B gets 5	Both get 1

These payoffs produce outcomes similar to many real-life situations provided certain conditions apply:

- The players interact an indefinite number of times so they do not know when the last interaction will take place.
- They remember how they both behaved in previous interactions.
- There is no way for the players to make enforceable threats or commitments.
- There is no way to be sure what the other party will do on the next move.
- There is no way to eliminate the other party or avoid the interaction.
- The players cannot change the payoffs.

The dilemma illustrated by the game is that whilst the group is better off if everyone cooperates, individuals can be better off by competing. However, faced with a party that competes, there is no benefit in continuing to cooperate. So competitive behaviour corrupts efforts at cooperating and causes everyone to compete. When this happens the group is worse off than if everyone had cooperated.

The results from a game of ten rounds of decisions by two parties based on the payoffs given above illustrate the point. If both parties cooperate on all ten rounds, they both score 30 making a total for the pair of 60. If one party competes and one cooperates on all ten rounds, the first party scores 50 and the second nil, making a total for the group of 50, which is less than when they both cooperated.

However, it is extremely unlikely that the second party would continue to cooperate and score no points faced with a relentlessly competitive approach from the first party. Assuming they decide to stop cooperating after three rounds and compete for the remaining seven, they score 7 and the opponent scores 22, making a total of 29. This is now significantly less than if they had both cooperated. The first player may feel good because they have won but they have less than if they had cooperated and the group is significantly worse off.

The total payoffs produced by ten rounds of the two players making consistent decisions are shown below.

		Decision by A	
		Cooperate	Compete
Decision by B	Cooperate	60	50
	Compete	50	20

The best group outcome results when both groups cooperate. The Bibliography includes reports of research that demonstrate that many natural and human situations fit this pattern of payoffs. These include construction projects using partnering to foster cooperative partnering. These ideas are more fully described in Robert Axelrod's very important book, *The Evolution of Co-operation*, listed in the Bibliography.

What did I say?

This game is played by small groups to help people understand the importance of listening carefully. One of the participants describes a current problem they are concerned about to the other members of the group. It helps if the facilitator can identify suitable people and problems beforehand. When the description is finished, one of the listeners restates the key features of the problem. Then another listener describes the emotions displayed by the original speaker. Then another listener describes the original speaker's motivation to find a solution to the problem. The original speaker comments on the accuracy and completeness of these statements and identifies any surprising insights.

The participants commenting on the original statement usually miss important points, invent features of the problem not mentioned by the original speaker and misunderstand key facts. The game can be repeated with people taking different roles to illustrate the need for open communication and discussion in building cooperative teamwork.

A wake-up call

A game that can be used to wake up participants is to ask them to list actions they could take to ensure that the project will fail. After five minutes of fun, the facilitator invites the participants to consider reversing each wreaking action to identify what is needed for the project to succeed. This may well identify important ideas that are worth including in the workshop outputs but the game serves its purpose if it causes people to become more animated.

Closing the workshop

- The workshop must agree what has been decided and this is helped by the facilitator keeping a running summary of decisions, which are displayed in the workshop room to give participants time to consider whether they agree with the decisions and the way they have been expressed.
- Participants are encouraged to suggest improvements to the wording so that over the two days several versions of the decisions may be on display.
- The final session reviews each major decision and the way it should be expressed so there is an agreed outcome.
- The workshop should agree the specific actions to be taken and the people responsible for taking those actions. These may be:
  - To continue the discussion of some issues on which agreement could not be reached
  - To hold another workshop to reconsider some key issues
  - To set up a task force to tackle a difficult problem
  - To undertake training or to review the membership of the project team.

20 Competent Teams

- Partnering depends on:
  - Teams being technically competent
  - Teams being experienced in their role in the project
  - Teams being effective
  - Teams working in cooperation with each other.
- Ensuring this requires partnering projects to have carefully designed selection procedures and training, coaching and mentoring may also be needed.

Technical competence

- People in construction are competent in many technologies, e.g. project managers, architects, engineers, piling specialists, bricklayers, steel erectors, cladding fixers, carpenters, electricians, painters, estimators, construction planners and construction managers.
- The broad range of technologies required by modern construction includes the techniques involved in financing, designing, planning and controlling work as well as its direct execution.
- An important trend in technological competence is multi-skilling, which means mastering a basic technology and knowing how to plan, control and improve its established methods and techniques. It may also include learning related technologies so teams can undertake broad tasks.
- First-line workers should take responsibility for helping establish their own objectives and targets.
- First-line workers should take responsibility for managing their own work including looking for ways of increasing their own efficiency and improving their products and services.

Level of work

- Work teams need to be experienced in dealing with the task complexity (which means the number of variables that have to be considered, how clearly they can be defined, and their rate of change) required by their role in the project organization.
- Work teams need to be experienced in handling information at the level of abstraction required by their role in the project organization.
- Work teams need to be experienced in making decisions that take account of the length of time taken for the outcomes to become apparent, associated with their role in the project organization. Low-level work deals with direct physical work in which outcomes are immediately apparent, while high-level work deals with abstract generalized ideas in which outcomes become apparent only after many years (Jaques, 1989).

Team roles

- Effective teams need at least one person able to play each of the following nine roles. One person may play several roles but all need to be provided by someone if a team is to be effective (Belbin, 1993).
  - *Coordinator* is the natural chairperson of the team. They are good at clarifying goals and ensuring the team agrees on priorities and reaches decisions.
  - *Shaper* presses for action, finds ways around obstacles and drives the team to action.
  - *Plant* is the creative source of original ideas and solutions to difficult problems.
  - *Monitor-evaluator* carefully dissects ideas the team is considering, weighs up the options and identifies problems.
  - *Resource-investigator* is the extrovert, enthusiastic, inspiring communicator who develops the external links that bring new contacts, ideas and developments into the team.
  - *Implementor* is the practical organizer who turns ideas into manageable tasks and then schedules and plans them.
  - *Team worker* holds the team together by supporting others, listening, encouraging and understanding.
  - *Finisher* checks details, worries about schedules and chases others with a sense of urgency that helps the team meet its deadlines.
  - *Specialist* ensures that the team is technologically competent.

Effective relationships

- See checklist 21.

21 Links Between Work Teams

Developing links

- Project organization structures that support partnering are shaped by the communication links people form and find useful, not by senior managers drawing organization charts.
- Effective communication links are part physical, part intellectual. They include:
  - Shared knowledge and experience about how, where and when the work of separate teams interacts.
  - Procedures for making joint decisions about all aspects of design, manufacture and construction.
  - Control systems that tell each work team how their work is contributing to overall project objectives.
  - Ways of resolving problems quickly in the interests of the whole project.
- As people communicate regularly, private languages develop so that much routine communication becomes virtually implicit and effort is concentrated on important matters.
- As people communicate regularly, shared cultures develop so that individual decisions take account of other people's interests.
- As people communicate regularly, misunderstandings become rare and communication becomes reliable and intended meanings are understood.
- Effective links develop when project teams and supply chains are kept together so that interactions are frequent.

Firms' responsibilities

- Partnering organizations should encourage work teams to interact and encourage the expectation that they will interact again in the future.
- Partnering organizations should avoid the problems that arise when there is a rapid turnover of staff so that people are not tempted to act for their own narrow advantage irrespective of the damage it does others because they will not be around to face the consequences.
- Partnering organizations should ensure that the structures and processes established by project teams ensure that cooperation is rewarded so that uncooperative behaviour is expensive compared with cooperating.
- Partnering organizations should invest in training that teaches people the values, knowledge and skills that promote cooperation. This includes helping people to recognize uncooperative behaviour when it occurs.
- Partnering organizations should take account of the guidance on internal partnering in checklist 23.
- Partnering organizations need to remember that cooperation is not a panacea; controls are needed to prevent collusive business practices that are not in the public interest.

Work team responsibilities

- Work teams involved in partnering take direct responsibility for organizing the links needed to work effectively. This is easiest when the selected work teams already have effective links but whatever their experience, teams should establish the links they need with other work teams.
- Work teams decide the links they need as they discuss how they will work together at partnering workshops, which should be designed to encourage teams to cooperate in establishing effective links.

Core team responsibilities

- The core team influences the patterns of communications that develop by the way they allocate work, decide its location and distribute resources.
- The core team influences the patterns of communications that develop by the way they set up information technology networks that connect some work teams and exclude others.
- The core team influences the patterns of communication that develop by the way meetings bring some work teams together and exclude others.
- The more the core team's decisions reflect and reinforce naturally occurring patterns of communication, the more effective they will be.
- The core team needs to ensure that the efficiency that comes from good communication links is devoted to achieving agreed work objectives.
- The core team should ensure that work teams are committed to agreed objectives, rewards reinforce objectives, and teams have feedback on their performance.
- The core team should ensure that full use is made of modern face-to-face and electronic links.

Face-to-face links

- Firms involved in partnering invest in making face-to-face communications more effective with the aim of fostering well-developed links between work teams.
- They check that meetings have clearly defined aims, bring together all the people affected by decisions and concentrate on finding practical, efficient answers.
- They use task forces of experts set up to find answers to serious, urgent problems.
- They use various forms of facilitated workshops to help project teams tackle important parts of their work.
- They set up common project offices to bring all the key decision-makers together.

- They invest in training and coaching to improve human interaction, time management, leadership and other facets of face-to-face communication.
- They fund social gatherings that range from an evening at the local pub to long weekends at tropical beach resorts.

Electronic links

- Firms involved in partnering invest in digital technology to help work teams work faster, more effectively and to higher standards.
- Information can now be where it is needed immediately and be totally up to date.
- Work teams can look at many options and simulate alternatives to measure the benefits, costs and risks of their decisions. Designs can be analysed to determine their safety, cost, value, environmental impact and buildability.
- Construction methods can be simulated to identify the safest and most effective ones.
- Work teams can remain in their own offices with all the support they provide yet electronic conferencing techniques give them many of the benefits of working in a common project office.
- Hot desking techniques allow people to work effectively as part of a team wherever they happen to be.

Cooperation

- Effective links between work teams are based on cooperation.
- Cooperation means helping others and not taking actions that damage others even if this means forgoing an immediate benefit.
- Stable cooperation is based on reciprocity so people expect others to reciprocate uncooperative behaviour as well as cooperative behaviour. This discourages them from starting trouble.
- Cooperation once established tends to grow because it provides more benefits than acting in ways that damage others and provokes them into reciprocating the adversarial behaviour.
- Cooperation is most likely to develop when work teams expect to work together in the future. When they do not have this expectation, it may be rational to act for their own narrow advantage irrespective of the damage it does to others.

Dealing with uncooperative behaviour

- The best strategy for everyone is never be the first to act uncooperatively. It is much more successful than any strategy that involves unprovoked, uncooperative acts.

- Uncooperative behaviour needs to be identified quickly and dealt with in time to deter the transgressor from continuing.
- There must be no doubt that uncooperative behaviour will be dealt with quickly and decisively. Letting it go unpunished merely encourages others to continue acting selfishly.
- Actions taken in response should be limited and it should be made clear that they will be discontinued as soon as the original transgressor returns to cooperative behaviour.

22 Task Forces

- A task force is a small group of individuals brought together to find an answer to a specific problem or to study a defined aspect of construction work.
- Task forces should have just one specific task, a relatively short time to complete the work, and sufficient resources to search widely for the best possible answer.
- Task forces should be disbanded as soon as they have agreed an answer to the specific issue they were asked to deal with.
- Task forces set up to develop performance improvements outside of any specific project usually include people drawn from firms involved in strategic arrangements who have directly relevant expertise. It often helps to add one or two external experts to guide the search into new areas.
- A task force can be an effective way of moving forward when a project team is stuck on a difficult problem. It allows the rest of the team, not forming part of the task force, to make progress in other areas of the project work.
- Task forces used during a project usually bring together a few members of the project team who have a real interest in finding a good answer to a specific problem. External experts may help guide the search into new areas but it is difficult for them to be useful when the task force faces project time pressures.
- Task forces are an important source of entirely new ideas.

23 Internal Partnering

This technical description will help clients, consultants, contractors and specialists involved in partnering to ensure that their own internal organization supports partnering.

- The most effective work teams in partnering projects have clear support from their parent firms for the decisions they make and the actions they take.
- The various actions needed to enable an organization to work cooperatively with others are sometimes called 'internal partnering'.
- Firms identify the key roles needed to lead and support the development of cooperative teamwork throughout the organization and select experienced people to be responsible for each.

Organization structures

- Managers are responsible for building links with customers and suppliers and supporting work teams and empowering them to do their best work rather than with internal control.
- Organization structures are encouraged to develop into self-organizing networks.
- Organization structures are streamlined with relatively few levels of managers and virtually no middle managers to encourage cooperative ways of working.
- Firms identify the key roles needed to develop links with key customers and suppliers and select an experienced manager to be responsible for each such link.
- Firms identify the key roles needed to develop links with other firms involved in strategic arrangements and select an experienced manager to be responsible for each such link.

Processes

- Encourage open communications by making all the information available to all parties and invest in communications and information technologies that help work teams improve their performance.
- Processes are streamlined so there are few duplicated actions, little checking and double-checking to encourage cooperative ways of working.
- Ensure that processes and commercial arrangements support and encourage efficiency and eliminate waste.

Culture

- Foster a cooperative internal culture, which creates an environment in which people are encouraged to achieve their potential.

- Encourage people to act on the basis that everyone is doing their best in the best interests of the whole organization. Discourage them from making assumptions about others' behaviour. Encourage them to ask questions that give others the opportunity to explain. Encourage everyone to be worthy of trust.
- Encourage everyone to take responsibility for their own actions, see mistakes as learning opportunities, and concentrate on finding answers without attempting to allocate blame.
- Deliberately break down suspicion and mistrust by emphasizing that everyone has a part to play in best-practice partnering.
- Focus on success by actively seeking to deliver the maximum benefits for everyone involved.

Training

- Empower everyone by education, training and induction that equip them with essential technical and partnering skills. Then give them the resources and authority to do their best possible work.
- Nurture an environment of continuous learning where best practice can be utilized, developed and shared by all.

External focus

- Firms give high priority to meeting the needs of their ultimate customers, the users and whoever provides the finance.
- Ensure that when other firms are employed, they are selected on the basis of achieving long-term sustainable value for money not the lowest price. The selected firms should be remunerated in ways that give them incentives to deliver good-quality construction on time and within budget.
- Agree common processes with other firms involved in partnering arrangements. This may mean replacing company methods and procedures with industry ones.
- Deliberately build long-term relationships committed to searching continuously for performance improvements.

Project support

- Ensure that teams are involved in projects early by assembling at the outset all those who have a major contribution to make to the current stage.
- Integrate design with construction process to ensure that designers take account of cost, time, quality, buildability, and the health and safety of those who construct, use, maintain and demolish the building or infrastructure.

- Set targets for quality, time, cost, health and safety, effectiveness of processes in terms of productivity and reduced waste, the quality of completed buildings or infrastructure including their fitness for purpose, total life-cycle costs and overall client and user satisfaction.
- Focus on outcomes in defining targets.
- Establish effective and reliable quality, time, cost and safety control systems and train staff to use them effectively.

Performance improvement

- Regularly measure the firm's overall and project performance to check that targets are being met, identify lessons and establish how performance has changed over time and how it compares with industry norms.
- Publicize case studies showing how partnering delivers greater benefits than traditional approaches for individuals, teams and firms.
- Invest in research and development aimed at achieving specific improvements in performance.
- Establish programmes of actions aimed at improving project performance over specified periods of between two and five years.

24 Controlling Quality in Partnering Projects

Make quality important

- Best practice quality is based on everyone involved in a project taking responsibility for quality.
- Best practice puts quality first in the trilogy of quality, time and cost.
- Everyone needs to recognize that there is no long-term future for them or their firms in producing poor quality quickly or cheaply.

Firms' responsibilities

- Partnering firms understand that getting work right first time is essential for high productivity and fast construction. Reworking, altering work already done, making good defects or simply having to return to a workplace to complete work left unfinished costs money, wastes time, destroys motivation and eats into profits.
- Partnering firms should aim to make zero defects a reality. The concept of practical completion, still included in many construction contracts, should be outlawed. Clients expect their new facilities to be handed over fully complete and see the idea of practical completion simply as an excuse for poor quality work. Reliable quality control can be used in construction and zero defects can be achieved.
- Partnering firms should foster a sense of total responsibility that does not rely on provisions in contracts but comes from a sense of pride in the work combined with a mature commercial awareness of the importance of happy clients.

Core teams' responsibilities

- The importance of quality should be emphasized by core team members at every opportunity.
- Core team members should ask questions about quality performance when they visit design offices, factories and the construction site. They should ensure that quality is discussed explicitly at every project meeting.
- Core team members should ensure that work teams can concentrate on doing their best work without being distracted by money issues or paperwork.
- The core team should ensure that contracts are based on terms which are fair to both parties so that conflicts do not arise about payments or claims for extra money.
- The core team should ensure that procedures allow everyone to take quality absolutely seriously and not spend hours dealing with paperwork.

- All work teams involved on site should attend a health and safety course that includes quality control procedures before being allowed on site.
- Quality should be linked to payments by defining completed work as including completing relevant quality procedures. This powerfully reinforces the importance of quality.

Designing for quality

- Quality begins with the use of tried and tested methods wherever possible so that the performance of details, components and systems is known and appropriate quality can be designed and planned into buildings.
- Appropriate quality means that a knowledgeable experienced choice is made of the right quality for each part of the end product.
- When a new solution is essential, it is properly researched. This may mean a research institute is called in to work with the design team to test and develop a new answer.
- Quality assurance begins with designers' own quality control systems. They define what is needed for quality design and provide a framework for defining what contractors' quality assurance systems should provide.
- Designers should undertake regular risk analyses of their designs to identify potential problems for the quality of the end product. In doing this they should consider any problems for health and safety during and after construction.

Constructing for quality

- Contractors' quality assurance plans should provide an overall quality plan for construction, general quality control procedures and specific procedures for each element and system.
- The core team should ensure that the contractors' procedures are sufficiently robust to allow them to rely on self-certification by the contractors.
- Self-certification should be supported by spot checks to ensure that quality control systems are being properly applied on site. These should include designers checking that their designs are interpreted correctly and that quality control systems are being applied correctly and that they are effective.
- On site, quality should be measured continuously by means of detailed, rigorous and carefully planned tests to ensure that standards never slip.
- Good practice is to require each work team to maintain a photographic record of tests and their results to provide a simple and comprehensive record of the application of quality control systems.

- Any quality problems should be investigated to identify the causes. Every necessary effort should be concentrated on finding a robust answer so the problem does not recur. This may mean altering a design, a construction method or a procedure. It may mean that some work teams need to be reminded of correct procedures or need additional training in quality procedures.
- Project teams should aim at ensuring that buildings work exactly as designed when they are handed over to the customer. Any defects that occur should be put right straight away.

25 Controlling Time in Partnering Projects

Make time important

- Time control is the responsibility of the whole project team.

The overall programme

- Best practice time-control systems are based on an overall programme that begins with the agreed completion date and works back to establish fixed milestones.
- Each milestone defines specific work that must be completed by the milestone date to ensure that the completion date is achieved.
- Milestones normally occur at monthly intervals and to ensure that projects finish exactly on time, programmes have slack built into them in various ways. (Work can sensibly be programmed on the basis of five normal working days each week, four weeks in each month and ten months in a year.)

Detailed programmes

- In the early stages of projects the definition of work required to meet a milestone needs to be flexible so that design decisions can be absorbed. As firmer decisions are made, the required work is defined more precisely.
- Each supply chain, and within that each work team, should undertake the detailed planning of their own work within the overall framework of milestones with the aim of defining a robust way of completing the work needed to meet the next milestone.
- Milestone plans should provide slightly more resources than will be necessary if the work goes very well and work teams should have the authority to use this flexibility in whatever ways they find necessary. Given this, there should be no flexibility over meeting milestones.
- Checking and coordinating the detailed programmes is undertaken by an experienced construction manager on behalf of the whole project team.
- Prior to each milestone the experienced construction manager produces a detailed programme of the agreed work needed to achieve the next milestone and everything that should be completed to achieve all the subsequent milestones.
- The detailed milestone programme identifies key deadlines for information from the client, the production of design information, any other information to be provided by one work team for another to work on, the manufacture of components and preparation on site.
- The detailed milestone programme provides part of the background information given to all those attending

milestone workshops. It should be reviewed and confirmed by the workshop, which provides a last opportunity to spot problems and deal with them before the programme is put into effect.

Week-to-week time control

- A progress meeting is held every week at which each work team reports one of the following situations:
  - Their work is on programme
  - They can deal with any problems they have
  - They need help from the rest of the project team.
- It is fundamental to partnering that this progress reporting is absolutely honest. Hiding problems should be regarded as a serious crime. In return for absolute honesty, time is not wasted trying to allocate blame for problems. The project team concentrates on finding the best possible answer that ensures they meet the next milestone.
- When problems arise the slack built into the overall programme may be used to bring work back onto programme. (Work can continue late into the night, seven days a week or a second shift can be worked by introducing extra work teams.)
- Following each progress meeting, the programme of work needed to meet the next milestone is updated by the experienced construction manager taking account of the decisions made by the progress meeting. This guides the next week's work.
- The whole project team should aim to keep the project exactly on programme or, when this proves impossible, to return it to the original programme as quickly as possible so that the project will be completed exactly on the agreed date.
- The core team should check that everyone involved concentrates on doing whatever is necessary to complete all the required work exactly on time.

Day-to-day time control

- The importance of sticking to the plan should be reinforced by beginning the working day with each team leader briefing their work team on the planned day's work. The briefing should describe the main processes, major deliveries, safety priorities, quality issues and any other points which need special attention so that every worker knows exactly what they are doing during the day and where it fits into the overall plan.
- Every team should continue its work until they have finished the agreed day's work.
- Day-to-day coordination and control is provided by a daily construction meeting attended by core team members and work team managers currently working on site held at a fixed time each day. Everyone focuses on solving problems quickly.

No discussion of contractual issues or claims for extra money or time is allowed. Peer pressure is directed at maintaining work exactly on programme with no question of relaxing quality standards. All positive suggestions are considered in turn until a reasonable consensus is reached. Then the meeting chairman announces the decision, which is recorded on a board in front of the meeting. Construction meetings should take 20 to 25 minutes and deal with the following:

- Problems that have arisen during the current day's work are discussed and resolved.
- The next day's work is agreed in accordance with the current overall programme.
- The meeting chairman reads out the agreed decisions to give everyone a second chance to spot any problems.
- Everyone is given a copy of the record of the agreed decisions.

26 Controlling Cost in Partnering Projects

The business case

- Best practice cost control begins with the client's business case, which defines the function, quality and value required in a new facility and therefore the maximum cost the client can afford to pay. The business case often states when the new facility needs to be available. This most directly influences time control but time may have a direct influence on value and costs.
- The client's business case establishes the main criteria to be met by the project and its overall budget.
- The feasibility of the criteria and budget should be tested by reference to similar projects and when necessary specific studies. These should include value and risk management studies to ensure that the project, on the particular site, given the circumstances facing the local construction industry, is feasible. The site should be thoroughly surveyed so that there are no surprises to undermine the project's feasibility.

Cost planning

- The initial framework of cost targets should require the team to make some defined improvements on previous best performance. The targets must be achievable and accepted as being achievable by the whole project team.
- A fixed sum that provides a reasonable profit and contribution to fixed overheads is agreed with each construction firm. This is guaranteed no matter what happens to other costs.
- A cost plan is produced that in total equals the client's overall budget for the project. The cost plan provides for all the agreed profits and overheads. The remaining money is allocated in as much detail as possible to the elements and systems needed to deliver the required function and quality on time.
- The cost plan should not include any contingency or risk allowance. This helps ensure that project decisions are made on the basis of good information and aim at ensuring that the customer gets full value for the agreed budget. Contingencies and risk allowances weaken the pressure to aim at providing the best possible value and offer no compensating advantages for good cost control.
- The project team undertakes value and risk management studies to ensure that the cost plan provides a robust basis for cost control.
- Some clients put limits on the cash flow available to finance the project. Any such restrictions must be taken into account in preparing the cost plan.

- The cost plan is progressively developed as further design decisions are made until there are well-defined targets for each element and system in the finished facility. Best practice is for each target to be agreed in cooperation with the firm responsible for the particular work. The targets should be based on everything going well so that none of them include contingency allowances. It may well be sensible to undertake value and risk management studies in respect of at least some of the main elements and systems.
- The project team holds a workshop to discuss the cost plan as soon as it is in a reasonably robust state. The workshop considers each of the targets, the assumptions on which they depend, the risks and any remaining uncertainties.
- At the end of the workshop, the whole team formally accepts joint responsibility for making the cost plan work so the budget is achieved and the customer gets everything they need from the project.
- As agreements are reached with the firms responsible for each element or system, their part of the cost plan can become more detailed.

Cost control

- As decisions are made the cost effects are monitored and recorded once a week in a detailed cost report that states the cost status of each element or system. In addition it highlights any threats to the target and any opportunities to make savings not already dealt with by the work teams directly involved.
- Cost reports are reviewed by the core team weekly and decisions made about all threats and opportunities so that the project delivers the best possible value for the customer within the budget.

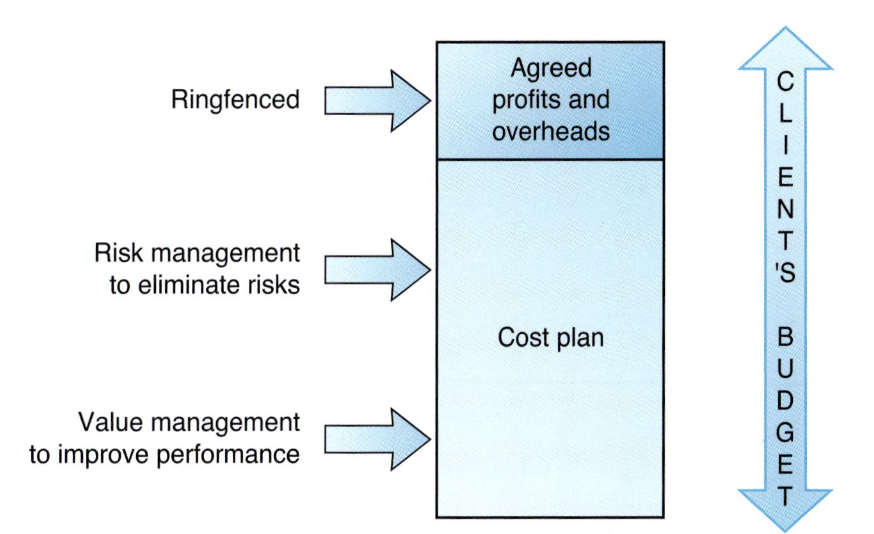

Figure 7.7 Cost-control process

- When difficult cost problems arise, it is the whole team's responsibility to search for savings to get the project back within budget. This may involve setting up a task force to find the best possible answer to some difficult problem. It may mean that some cost targets have to be cut and different designs produced. Best practice does all this without ever considering reducing the firm's profit and fixed overhead margins or giving the customer worse value.
- Best practice challenges project teams to be creative in finding the best possible answers within the client's overall budget and their own fixed allocation of profit and overheads.
- The project team should keep in mind the broad outline of the cost-control process shown in Figure 7.7.

27 Project Execution Plan

This checklist identifies the issues that should be considered by clients and projects core-teams in producing the project execution plan.

The client

- Client's objectives including the business plan insofar as it affects or is influenced by the project.
- Client's representatives and their authority.

The project team

- Selection processes for project team members.
- Core team members and their contractual responsibilities.
- Consultants and their contractual responsibilities.
- Lead firms in each supply chain and their contractual responsibilities.
- Contractors and suppliers and their contractual responsibilities.
- Third parties likely to be involved.

Risks

- Risk assessment including particularly a careful evaluation of the site, ground conditions and access.
- Indemnities, insurance and warranties.

Agreed objectives and decision-making

- Mutual objectives emphasizing what each firm involved should get from the project.
- Decision-making systems including procedures, standards, constraints, targets, control systems, meetings, information systems and feedback systems.
- Problem-resolution procedures.
- Performance improvements including a summary of how they are intended to be achieved.

Design

- Design concepts and scheme design.
- Detailed design.
- Design coordination including the assessment and control of design changes.

Control systems

- Quality assurance systems and the resulting records of inspections and tests.
- Agreed completion date and overall programme including milestones.
- Time-control systems and progress reporting.

- Agreed budget and overall cost plan.
- Cost-control systems and cost reporting.

Construction and handover

- Construction method and coordination including the assessment and control of construction changes.
- Commissioning and handover.
- Operation of the completed facility.
- Safety, health and environmental issues including actions to ensure compliance with the construction design and management regulations.

Project feedback

- Post project evaluation and feedback to future projects.

The project execution plan should be openly available throughout the project organization. This is most likely to be achieved by maintaining it in an electronic form supported by a state-of-the-art search engine. The project execution plan should be kept absolutely up to date and formally reviewed by the core team as each milestone is reached.

28 Project Handbook

CIOB's *Code of Practice for Project Management for Construction and Development* provides a detailed checklist for the contents of the project handbook. This is not repeated here because it deals with the same issues as the project execution plan but with different objectives. The objectives of the two documents are:

- The project execution plan gives the client and organizations providing finance a statement of the stages and processes of the project demonstrating quality, financial and time control and an agreed means of achieving the client's objectives.
- The project handbook guides the project team in the performance of its duties which are the design, construction and completion of a project that meets the client's objectives for the performance and quality of the new facility within the agreed budget and completion date.

29 Value Management

Value for money

- Value management aims to establish which features of a new building or infrastructure deliver value to the client and identify ways of maximizing these relative to the cost the client is prepared to finance.
- Value management is carried out at workshops held at significant stages of projects.
- Value management studies can contribute to the following:
  - Statement of the client's objectives
  - Overall concept and feasibility study
  - Selection of project team
  - Formal brief
  - Scheme design and reviews of its quality, time and cost
  - Detail design and reviews of its quality, time and cost
  - Construction method and reviews of its quality, time and cost.

Value management objectives

- A value management study begins by identifying the features of the project that will be considered. A careful review of the following issues can produce ideas that help in deciding the broad objectives for a value management workshop.
  - Client's business plan including the potential for more efficient operations
  - Aesthetic design including the impact on users, customers and neighbours
  - Use of prefabricated elements and system components
  - Flexible design including future developments and eventual sale of the facility
  - Internal comfort conditions including the benefits of natural lighting and ventilation and of sophisticated control over internal conditions
  - Certainty of quality, time and cost
  - Quality including the benefits of zero defects
  - Completion date including the benefits of fast construction
  - Capital and whole-life costs including taxation issues
  - Environmental and sustainability issues.
- The broad objectives are considered to agree specific objectives, which should be expressed as measurable targets. The specific objectives need to relate to the current stage of the project. The following are typical objectives for value management studies.
  - Confirm that a new facility is necessary and there is no way of meeting the client's business objectives that delivers more added value.

- Define the overall concept in a way that is feasible and provides a brief that will deliver a greater net addition to value than would result from the current overall concept and brief.
- Review the main options considered in producing the scheme design to identify answers that deliver a greater net addition to value than would result from the current design.
- Consider the way the client intends using the new facility to identify ways of improving the scheme design and making the client's business more efficient.
- Review specific detail designs to look for ways of improving performance in terms of quality, time or cost consistent with making a net addition to value.
- Review the construction methods to look for ways of improving performance in terms of quality, time or cost consistent with making a net addition to value.

Preparing for a workshop

- A workshop facilitator, sometimes called a study leader, should be appointed and the people to attend the workshop should be identified. They should include everyone with a real interest in the outcome who is likely to make a significant contribution.
- At this point it is good practice to hold an initial meeting of five or six key people involved to ensure that the objectives and targets are properly understood and fully reflect what the project needs. This meeting should agree the information needed for the workshop to be successful and how it will be produced. It also checks and agrees the list of those to be invited to attend and the duration and timing of the workshop.
- The information produced prior to the workshop should include the current descriptions of the features of the new facility that deliver value to the customer. These are sometimes called the principle functions or value drivers. The information should define how the targets for improving the value will be measured.
- A suitable venue should be selected to ensure that participants can concentrate on the workshop.
- Any participants not familiar with value management workshops should be given training in the process and techniques likely to be used to enable them to contribute effectively.
- The participants should be sent a formal invitation well before the workshop that tells them the objectives and targets, provides the background information needed at the workshop and explains all the practical arrangements.

The value management workshop

- The actual workshop reviews the value being delivered by current proposals for the project and the likely cost.
- Various techniques are used to search for ways of increasing the value relative to the cost that are consistent with the client's objectives.
- The best ideas for adding value identified in this way are selected on the basis of their contribution to achieving the workshop's targets. These ideas are discussed further to establish their feasibility and check their likely contribution to net value. Most ideas need further development.
- The workshop defines what actions need to be taken and who is responsible for carrying them out.
- A formal description of all the ideas the workshop decides should be taken further should be produced.

Actions following the workshop

- The people given responsibility for developing workshop ideas into detailed proposals should report the results to the core team no more than two weeks after the workshop.
- Each report should include detailed actions that have been discussed with the work teams involved. It should also provide an evaluation of the proposal in terms of the workshop targets and its effects on value and cost.
- The core team decides what is done with each of the developed ideas and produces a short report of the added value provided by the ideas generated by the value management workshop.
- The core team's report is distributed throughout the project team to focus attention on the need for everyone to consider ways of adding net value, to help establish the benefits of value management and to congratulate those involved in the value management workshop.

Maintaining the focus on value

- Between workshops everyone making decisions about a project should take account of the likely effects on value and cost. The core team should regularly check that this is happening.

30 Risk Management

Risk in construction projects

- Construction always involves some risks. Whether they have a negative or positive impact depends on when they are identified and how they are managed.
- Clients should be proactive in insisting that risks are managed right from the start of a project because many unmanaged risks tend to fall back on them in one way or another.
- Risk management aims to reduce uncertainties about the outcomes of the project and can contribute to the following:
  - Project strategy
  - Overall concept
  - Project organization including procurement
  - Scheme design
  - Detail design
  - Construction methods.

General principles of risk management

- Risk management should be started at the earliest possible stage of construction projects because good decisions in the early stages often avert the biggest and most expensive risks.
- The earlier the whole project team is appointed the better the risk management process will be. Contractors, consultants and other key suppliers bring knowledge and experience of construction, delivery and related financial risks that are helpful in managing risks.
- Risk management workshops should be held at significant stages of projects to identify and evaluate all risks and make robust decisions about how they are managed.
- Risk management workshops should be directly responsible to the project core team.

Preparing for a workshop

- A workshop facilitator, sometimes called a study leader, should be appointed and the people to attend the workshop should be identified. They should include everyone with a real interest in the outcome who is likely to make a significant contribution.
- Prior to the workshop the people invited to attend should ensure they understand the project in depth.
- The workshop facilitator should interview senior managers in the main organizations involved in the workshop to obtain robust information.

- The information they need depends on the stage reached but in general describes the objectives, benefits, main stakeholders' interests, physical and business environments, financial and time constraints, design, quality, time and cost.
- In assembling this information, the workshop facilitator should ask the people interviewed to identify the main risks facing the client's organization. Risks may arise in any aspect of the client's business including the following:
  - Commercial and financial
  - Demand and competition
  - Operating and staffing
  - Technological innovation
  - Political and legislation.
- The initial information-gathering should also identify the main risks facing the project team. These may arise from any aspect of the project including the following:
  - Client's objectives
  - Impact of the project on the local economy
  - Impact of the project on the local community
  - Planning and other regulations
  - Interventions by third parties
  - Financial restrictions imposed by the client
  - Restrictions imposed by insurance or other financial institutions
  - Site characteristics and location including the weather
  - Access to the site
  - Market conditions in the local construction industry
  - Availability of specialized resources
  - Influence of trade unions
  - Project organization and procurement
  - Design including the need to find new answers
  - Design information, its volume, timing and distribution
  - Quality control and the potential for defects
  - Time control including information flows and decision-making
  - Cost targets and control systems
  - Construction methods including the need to find new answers
  - Timely appointment of project team members
  - Competence of project team members
  - Adequacy of resources devoted to project
  - Decision-making systems
  - Natural or man-made disasters.
- Objective and verifiable information about the site, ground conditions, access and if appropriate existing property should be prepared to guide or test early decisions. This is a vital first step in the active management of risks.

Partnering and risk management

- When a client is considering using partnering, the following questions should be answered as part of an initial risk management study.
  - What kind of relationship is suggested by the business case for the project?
  - Would partnering be appropriate and if so, what exactly is the case for using this approach?
  - Is our organization ready to work with consultants, contractors and specialists on a partnering basis?
  - Do we have the leadership, skills and capability to use partnering effectively?
  - What is our track record in partnering with consultants, contractors and specialists?
  - Could existing relationships, ours or those of other organizations, act as a model for using partnering on the project?
  - Can we define success in establishing a partnering relationship and set targets, milestones and measures that will enable us to assess how successful we have been?
  - Assuming a partnering relationship can be established, will users and stakeholders 'sign up' to it and add momentum to its development?
  - What kind of construction firm could manage the risks we envisage allocating to them?
  - Is it realistic to expect a construction firm to take on these risks and can we give them sufficient control so that they can manage them?
  - How will the local construction market react to a partnering approach to the project?

The workshop

- The workshop takes the form of a well-run meeting chaired by the workshop facilitator.
- It begins by identifying the main risks. Various techniques exist to help the search and detailed descriptions of effective techniques can be found in the publications listed in the Bibliography. Also the *Code of Practice for Project Management for Construction and Development* includes a good checklist of possible risks.
- As risks are identified, they are listed in a risk register. The risk register forms the basis for risk management. It comprises a description of each risk, states the probability of it occurring, its impact on value, quality, time and cost, how the risk will be mitigated and who is responsible for taking the mitigating actions.
- The workshop makes rough evaluations of the probability that each risk will occur and the impacts. In many cases this information will be produced before the workshop and be robust. In other cases the risk will need to be analysed in

detail after the workshop before the risk can be quantified with any useful degree of confidence.

- Quantifying risks is an imprecise activity that combines objective measures and subjective judgments. It often helps to produce a range of estimates of the likelihood of the risk occurring and its impacts. The risk management publications listed in the Bibliography include descriptions of effective ways of producing and using these estimates and warnings about the limitations of the techniques.
- The workshop's preliminary evaluation serves to identify the most significant risks. The workshop then concentrates on deciding how they should be dealt with. There are four broad strategies for dealing with risks:
  - Eliminate the risk
  - Reduce the risk
  - Take out insurance
  - Accept the risk.
- Each risk should be considered and one of the four strategies adopted. The specific actions resulting from these various strategies may include changing the client's objectives, overall concept or brief. They may mean redoing the feasibility study. They may include surveys or site investigations. They may mean altering the design. They may mean changing the composition of the project team. They may mean examining the feasibility of developing new technologies, processes or methods. They may mean using different construction methods. They are very likely to include some reallocation or at least some clarification of who carries which risks. In extreme circumstances, the best action may be to abandon the project.
- The workshop is likely to have time to agree specific strategies for only the largest risks that threaten a cancellation or major failure of the project. Others can be considered outside of the workshop. It is important to identify who is responsible for dealing with every risk listed in the risk register.

Ongoing risk management

- It is sensible to consider whether any of the major risk factors may change and if so they may need to be monitored throughout the project.
- Throughout projects, anyone identifying a risk should check that it is listed in the risk register and if not ensure that the core team are told so they can consider how it should be dealt with.
- Each successive risk management workshop begins by considering whether the risk register is complete. Then it concentrates on identifying new risks associated with the stage reached and deciding how they will be dealt with.
- The core team reviews the risk register at their progress meetings to ensure it is up to date and all the risks are being dealt with.

31 Benchmarking

Performance improvement

- Benchmarking is a tried and tested process that uses carefully devised targets to continuously improve performance.
- Benchmarking is based on researching the extent to which others carry out the same or similar processes more efficiently, identifying how they do this and selecting features of their methods that can be used to improve performance. The improvements are then put into practice and the results measured.
- Benchmarking is an on-going process and so the measured results provide a starting point for the next cycle of improvement.
- Benchmarking makes construction teams aware of best practice and helps them understand how they can apply its key features to their own work. In this way benchmarking engages people in a systematic search for ways to improve their own performance.

The benchmarking process

- Benchmarking consists of a series of practical techniques and the Bibliography includes guides to benchmarking tailored to the needs and circumstances of the UK construction industry. Figure 7.8 shows the distinct steps that need to be taking in using benchmarking.

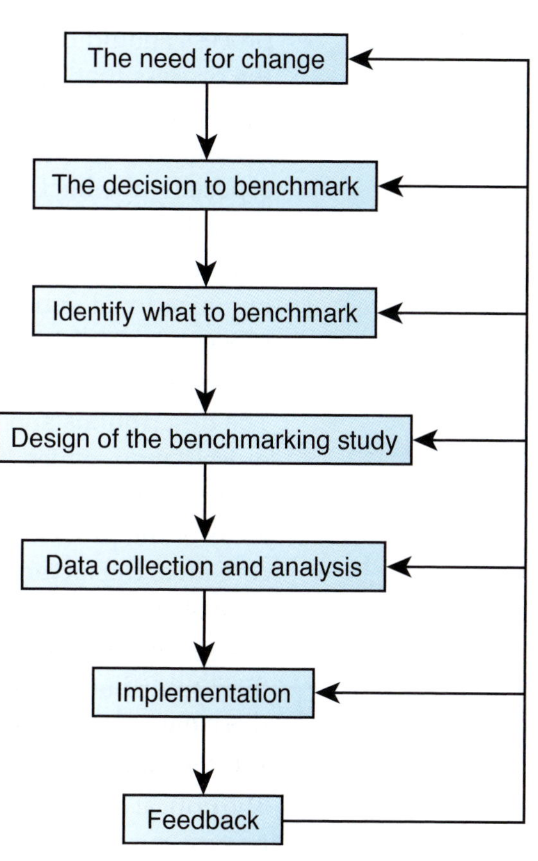

Figure 7.8 Benchmarking process

The need for change

- The need for benchmarking arises when an organization recognizes that it needs to change to improve performance. This may be triggered by: reports describing the performance achieved by others; Constructing Excellence in the Built Environment's key performance indicators; internal concerns over profit levels; the share price falling; or threats from competitors.
- The need to improve performance should be discussed at an internal workshop. The aims are to consider whether there really is a need to improve performance, define the problem that gives rise to this situation, identify the causes of the problem and suggest what needs to be changed. The output is an agreed statement describing the need for change, which provides the first step in benchmarking.

The decision to benchmark

- The second step is the decision to use benchmarking. There are costs in using benchmarking including training people in the process of benchmarking, workshops and facilitators. The benefits can far outweigh these costs; nevertheless they need to be considered in deciding to benchmark.

Identify what to benchmark

- The third step is identifying what to benchmark. The agreed need for change statement will help to identify which process should be benchmarked. It is sensible to start with the process that appears to provide the greatest need for improvement and the biggest benefits if a better answer is found. The ease with which change can be introduced, especially the likely attitudes of the people involved, should also be taken into account.
- The next step is to assemble existing descriptions of the organizational structures and methods used in the selected process, including the strategic measures of performance currently used by management.
- The descriptions and measures provide a starting point for a process map, which is prepared at the workshop by the staff responsible for the process. The aim is to identify the core areas of the process that add value for customers. For each core area, the factors critical to its success are identified by discussion at the workshop.
- The success factors should be prioritized on the basis of their contribution to the organization's overall success. The workshop should agree how the top priority success factors should be measured.
- The critical success factors and measures, together with the process maps of the targeted areas, form the preliminary framework for a benchmarking study.

Design of the benchmarking study

- The next main step is to design the benchmarking study. Benchmarking compares strategic measures from different organizations to identify differences in performance. The other organizations may be other divisions of the organization, competitors or non-competitor organizations undertaking the same function or generically the same activity. It is sensible to identify five or six benchmarking partners that use a range of approaches. Partners have to be convinced that there is a real probability of mutual benefits in undertaking a benchmarking study.
- The benchmarking partners' level of commitment becomes apparent as the details of the study are agreed. A critical stage often arises when the partners attempt to agree the detailed measurements to be used in comparing their performance. These must be defined and agreed by all the parties. This includes defining the precise data needed, the collection methods, the role and responsibilities of each partner and the degree of confidentiality that applies. Putting these agreements into practice and collecting the data needed to compare performance gives rise to significant work by all the parties. Agreeing the time and resources required to carry out this stage of the benchmarking study provides a real test of partners' commitment.
- Having agreed these key details and maintained the commitment of all the parties, the staff likely to be affected by changes aimed at improved performance should be told what is being done. They should have an opportunity to comment on the arrangements and to be involved in the study.

Data collection and analysis

- The next main step is the actual data collection and analysis. The rules of measurement should be tested in a pilot study to ensure that they are practical and provide robust measures of performance that are genuinely comparable.
- Having completed a pilot study, the measures and data collection methods to be used should be reconfirmed or altered to take account of the results. Then as the performance of all the partners is measured, regular checks should be made that consistent methods are being used and the measurements are truly comparable.
- When the data is assembled and checked carefully for ambiguities and gaps, it should be described in a formal written report that is available to all the parties involved in the study.
- Then the data needs to be analysed to identify differences in performance between the partners. Significant differences should be investigated to ensure that the data is accurate. Then the reasons for these differences must be identified by comparing the organization's own process analyses with similar descriptions of the best performing organizations.

- Each significant performance gap should be reviewed to decide if it is critical to the organization's success. If it is, then targets and timetables should be set for closing the gap in those organizations displaying weaker performance. Targets that will give quick wins help establish the benefits of benchmarking. The actions needed to achieve the target should be planned using the process analyses produced during the study.

Implementation and feedback

- The detailed targets and plans should be made available to all the parties involved in a formal report that is communicated to those whose support will be needed and those who will be affected by changes.
- Then the planned actions are put into effect. They should provide for feedback and regular progress reviews at which the parties discuss successes and failures. Shortfalls should be investigated and actions taken to achieve the targeted improvements.
- The improved performance level achieved as a result of the benchmarking study should be embedded in a new target or benchmark, which is established as the organization's new way of working.

32 Partnering Health Check

This technical description helps partnering organizations check how effectively partnering is being used.

- The following statements and issues should be discussed at interviews with a cross-section of project team members. Each statement should be allocated a score as follows:
  1. unsatisfactory, requiring a major review
  2. problems exist that need to be dealt with urgently
  3. making progress but more is needed
  4. well established
  5. excellent
- The scores in each of four main categories should be expressed as a percentage to give a health check score for people, teamwork, processes and outputs. All the scores should be totalled to provide an overall percentage which is the project's health check score.
- Scores should be related to previous health checks on the project, targets agreed for the project, and scores from similar projects.
- Individual questions or issues scoring more than 10% below the overall average should give rise to a report for the core team describing comments made during the interviews that help explain the poor performance.
- The following statements provide the basis for the partnering health check.

People

I know all the members of my work team.

I know the team leaders of all the work teams we work with.

I know all the members of the core team.

People are open and honest.

People understand others' problems.

People cooperate.

People support each other.

People act as team players.

People do not let their own interests damage the project.

People enjoy working on this project.

I am being paid fairly for my work on the project.

Teamwork

People are fully involved in the project in time to make a full contribution.

Information is communicated openly.

People concentrate on solving problems without allocating blame.

People provide honest feedback.

People actively encourage and help each other.

The project team is a real team.

The core team provides real leadership.

The core team allows work teams to make their own decisions.

My work team works well with other work teams.

My work team is a real team.

Processes

I understand how decisions are made on the project.

Design information is available on time.

The design is not changed unless it is in the best interests of the project.

Everyone knows their own work programme for this week and next week.

Quality control is fair and effective.

Cost is not a driving force causing corners to be cut.

We have time to do our work properly.

Time is not wasted on this project.

We have sufficient resources to do our work properly.

Resources are not wasted on this project.

Feedback is accurate and in time to help in making decisions.

We all have feedback on our own work team's performance.

Meetings are used to make good decisions.

Ideas for improving our work are encouraged and taken seriously.

People who suggest good ideas are rewarded sensibly.

Outputs

The project is achieving zero defects.

The construction site is safe.

The project is on time.

The project is within its budget.

Progress is good and effective.

Problems are solved quickly.

The project is not disrupted by problems.

The project is achieving its target performance improvement.

The project is the best one I have ever worked on.

The new facility will make a significant contribution to the client's business.

The client is happy with the new facility.

The new facility will make a significant contribution to the local community.

Our firm is making a fair profit on the project.

33 Teams in Construction Organizations

Various kinds of teams are referred to throughout this code of practice and the following glossary explains their roles.

A team comprises a number of individuals organized to work together. In this code of practice the term includes the plant and equipment needed for the team to do effective work.

As a team works together long term they adopt cooperative teamwork as described in Section 4.10 and become a mature team, which if well motivated achieves high levels of performance in all aspects of its work.

Audit team

A team responsible for auditing financial accounts.

Client's internal team

A team established by the client to ensure that the client's interests are properly taken into account by the project team that they appoint to undertake a construction project; it usually comprises members of the client's staff plus consultants as described in Section 3.2.

Competent team

The requirements for a team to be considered competent are given in checklist 20.

Construction team

A team that undertakes any kind of construction work, e.g. setting objectives, designing, planning, controlling, manufacturing, bricklaying, erecting steelwork, installing heating systems, etc. In the context of this code of practice it is an alternative term for team and work team.

Consultant team

A team provided by a consultant that undertakes specialist work.

Cooperative team

A team that is competent and has cooperative relationships internally between the individuals that comprise the team and externally with the other teams it needs to interact with.

Core design team

A team which is the part of the design team that provides overall direction and leadership for the design of an individual construction project.

Core team

A team which is the part of a project team that provides overall direction and leadership for the project as described in Section 3.9.

Design and construction team

An alternative term for project team.

Design team

A team responsible for the design of an individual construction project; it normally comprises many work teams including architects, engineers and specialists.

Effective team

A team that has developed through the stages described in Section 4.10 so they work on the basis of cooperative teamwork.

Facilities management team

A team responsible for the operation and maintenance of a building or other constructed facility.

Framework team

A team with overall responsibility for organizing a framework arrangement on behalf of a client organization.

Integrated team

A team in which the individuals that comprise it communicate and work together effectively.

Interface team

A team set up by a group of firms that are partnering over a series of construction projects to deal with issues arising in respect of a specific joint activity, e.g. design, technology, quality, time, cost, safety or other equally important issues arising from their joint partnering activities as described in Section 6.12.

Internal partnering team

A team internal to one firm that provides direction and leadership for the firm's use of partnering as described in Sections 5.2 and 6.12.

Internal team

A shortened version of the client's internal team or internal partnering team depending on the context.

Lead team

A team within a supply chain team that provides the overall direction and leadership for the supply chain and represents the interests of the supply chain team in dealing with the rest of a project team and the client's internal team.

Legal team

A team responsible for undertaking legal work.

Management team

A team responsible for undertaking the management responsibilities for an individual construction project; it normally comprises many work teams including construction planners, estimators, buyers and site supervisors.

Multi-skilled team

A team competent in technical specialisms and a range of related skills as described in Section 5.8.

Partnering team

A team using partnering to help it undertake its work more efficiently.

Project team

A team responsible for an individual construction project; it normally comprises many work teams. It could more accurately be called a project organization but the term project team is well established in the construction literature and is used in this code of practice.

Selection team

A team set up to select firms to provide work teams that will form part of a project team; it is normally set up by the client's internal team.

Specialist work team

A team that undertakes specialist construction work to provide a physical element or system of the constructed facility; it normally comprises many work teams including those responsible for designing, manufacturing, assembling on site, etc.

Strategic team

A team set up by a group of firms that are partnering over a series of construction projects to provide direction and leadership for their joint partnering organization as described in Sections 6.4 and 6.12.

Supply chain team

A team responsible for a complete supply chain; it normally comprises many work teams including those responsible for designing, manufacturing, assembling on site, etc.

Task force

A team set up to undertake a specific task within a specific (usually short) timescale as described in checklist 22.

User team

A team that represents the interests of those who use a building or other constructed facility to undertake the work or other activities that the facility is designed to accommodate.

Work team

A team that undertakes any kind of construction work, e.g. setting objectives, designing, planning, controlling, manufacturing, bricklaying, erecting steelwork, installing heating systems, etc. In the context of this code of practice it is an alternative term for team and construction team.

Bibliography

The publications listed in the Bibliography are selected to provide a short list of further reading to give readers access to the background to the advice given in this book.

Cooperation and teamwork

Axelrod, R. (1990) *The Evolution of Co-operation*. Penguin, London.

Belbin, R.M. (1993) *Team Roles at Work*. Butterworth-Heinemann, Oxford.

Bennett, J. (1997) *Measuring the Benefits of Long-Term Relationships in the UK Building Industry*. Final Report: EPSRC ref GR/J64474, Swindon.

Bennett, J. (2000) *Construction – The Third Way*. Butterworth-Heinemann, Oxford.

Capra, F. (1996) *The Web of Life*. HarperCollins, London.

Carlisle, J.A. and Parker, R.C. (1989) *Beyond Negotiation: Redeeming Customer – Supplier Relationships*. Wiley, Chichester.

Covey, S.R. (1989) *The Seven Habits of Highly Effective People*. Simon and Schuster, New York.

Hutton, W. (2002) *The World We're In*. Abacus, London.

Seabright, P. (2004) *The Company of Strangers*. Princeton University Press.

Womack, J.P., Jones, D.T. and Roos, D. (1990) *The Machine that Changed the World*. Macmillan, New York.

Management techniques used in partnering

CIOB (2002) *Code of Practice for Project Management for Construction and Development*. Blackwell, Oxford.

Constructing Excellence (2004) *Demonstrating Excellence: An Evolution of the Programme of Demonstrations*. Constructing Excellence, London.

Dallas, M. (2005) *Value and Risk Management: A Guide to Best Practice*. Blackwell, Oxford.

Fisher, R. and Ury, W. (1981) *Getting to Yes*. Hutchinson Business, London.

Galbraith, J. (1973) *Designing Complex Organizations*. Addison-Wesley, Reading, MA.

Green, S.D. (1992) *A SMART Methodology for Value Management*. CIOB Occasional Paper 53.

Imai, M. (1986) *Kaizen: The Key to Japan's Competitive Success*. McGraw-Hill, New York.

Jaques, E. (1989) *Requisite Organization: The CEO's Guide to Creative Structure and Leadership*. Casson Hall, Arlington, VA.

Lock, D. (ed.) (1994) *Gower Handbook of Project Management.* Gower, Aldershot, UK.

Morgan, G. (1986) *Images of Organization.* Sage, London.

Pickrell, S., Garnet, N. and Baldwin, J. (1997) *Measuring Up: A Practical Guide to Benchmarking in Construction.* Construction Research Publications, London.

Partnering in Construction

Bennett, J. and Baird, A. (2001) *NEC and Partnering: The Guide to Winning Teams.* Thomas Telford, London.

Bennett, J. and Jayes, S. (1995) *Trusting the Team: the Best Practice Guide to Partnering in Construction.* Reading Construction Forum, Reading, available from Thomas Telford, London.

Bennett, J. and Jayes, S. (1998) *The Seven Pillars of Partnering.* Reading Construction Forum, Reading, available from Thomas Telford, London.

Construction Industry Board (1997) *Partnering in the Team.* Thomas Telford, London.

Construction Industry Council (2002) *A Guide to Project Team Partnering.* Construction Industry Council, London.

Green, S.D. (1998) The technocratic totalitarianism of construction process improvement: a critical perspective. *Engineering, Construction and Architectural Management,* **5**(4), 376–86.

Green, S.D. (1999) Partnering: the propaganda of corporatism. *Journal of Construction Procurement,* **5**, 177–87.

Holti, R. and Standing H. (1996) *Partnering as Inter-related Technical and Organisational Change.* Tavistock, London.

Holti, R., Nicolini, D. and Smalley, M. (2000) *The Handbook of Supply Chain Management.* CIRIA, London.

Loraine, R. and Williams, I. (1997) *Partnering in the Public Sector.* European Construction Institute, Loughborough, UK.

Ogunlana, S. (ed.) (1999) *Profitable Partnering in Construction Procurement.* Spon, London.

Rackham, N., Friedman, L. and Ruff, R. (1996) *Getting Partnering Right: How Market Leaders are Creating Long-term Competitive Advantage.* McGraw-Hill, New York.

Ring, P. S. (1997). Processes facilitating reliance on trust in inter-organisational networks. In M. Ebers (ed.) *The Formation of Interorganisational Networks,* Oxford University Press.

Roe, S. and Jenkins, J. (2003) *Partnering and Alliancing in Construction Projects.* Sweet & Maxwell, London.

Guides to best practice

Bennett, J., Pothecary, E. and Robinson, G. (1996) *Designing and Building a World-Class Industry*. Centre for Strategic Studies in Construction, Reading.

Comptroller and Auditor General (2001) *Modernising Construction*. The Stationery Office, London.

Construction Task Force (1998) *Rethinking Construction*. DETR, London (The Egan Report).

Latham, M. (1994) *Constructing the Team*. HMSO, London.

Selecting Construction Firms

CIOB's Chartered Building Company Scheme (www.cbcscheme.org.uk).

Construction Industry Board (1996) *Selecting Consultants for the Team: Balancing Quality and Price*. Thomas Telford, London.

Construction Industry Board (1997) *Code of Practice for the Selection of Subcontractors*. Thomas Telford, London.

Construction Industry Board (1997) *Code of Practice for the Selection of Main Contractors*. Thomas Telford, London.

Construction Line (www.constructionline.co.uk).